Microstrip Circuits

WILEY SERIES IN MICROWAVE AND OPTICAL ENGINEERING

KAI CHANG, Editor
Texas A & M University

INTRODUCTION TO ELECTROMAGNETIC COMPATIBILITY
Clayton R. Paul

OPTICAL COMPUTING: AN INTRODUCTION
Mohammad A. Karim and Abdul Abad S. Awwal

COMPUTATIONAL METHODS FOR ELECTROMAGNETICS
AND MICROWAVES
Richard C. Booton, Jr.

FIBER-OPTIC COMMUNICATION SYSTEMS
Govind P. Agrawal

OPTICAL SIGNAL PROCESSING, COMPUTING AND NEURAL NETWORKS
Francis T. S. Yu and Suganda Jutamulia

MULTICONDUCTOR TRANSMISSION-LINE STRUCTURES:
MODAL ANALYSIS TECHNIQUES
J. A. Brandão Faria

MICROSTRIP CIRCUITS
Fred Gardiol

MICROWAVE DEVICES, CIRCUITS, AND THEIR INTERACTION
Charles A. Lee and G. Conrad Dalman

Microstrip Circuits

FRED GARDIOL
École Polytechnique Fédérale de Lausanne
Switzerland

A WILEY-INTERSCIENCE PUBLICATION

JOHN WILEY & SONS, INC.

NEW YORK / CHICHESTER / BRISBANE / TORONTO / SINGAPORE

Library of Congress Cataloging in Publication Data:
Gardiol, Fred E.
     Microstrip circuits / Fred Gardiol.
        p.   cm. —(Wiley series in microwave and optical engineering)
     "A Wiley-Interscience publication."
     ISBN 0-471-52850-1
     1. Microwave circuits.   2. Strip transmission lines.   3. Microwave
integrated circuits.   I. Title.   II. Series.

TK7876.G38   1994
621.381'32—dc20   93-19436
                                                                          CIP

Preface

BASIC DESCRIPTION

The field of *microwaves* extends over the gigahertz region of the electromagnetic spectrum and was traditionally used for radar and communications, in particular with satellites and faraway space probes. In recent years, the field's scope has considerably widened, because transistors operate at increasingly high switching rates, bringing computer signals well within the microwave range.

Traditionally, microwaves used to be the domain of waveguides, mysterious hollow metallic tubes through which electromagnetic waves get transmitted and reflected along the boundaries. Circuits made with such tubes (plumbing) were heavy and cumbersome, taking up a lot of space and giving rise to some interesting topological problems, solved by means of bends and twists.

This situation changed considerably with the advent of printed circuits, the ones more commonly used microwaves being microstrips. The extension of the photolithographic process to the microwave range let designers reduce size, weight, and cost of low-power components and systems. The process is well suited for series production of circuits and antennas, since lumped circuit and active devices can easily be connected by sections of transmission line.

However, there is a significant difference between low-frequency printed circuits and their microwave counterparts: at microwave frequencies, all dimensions become important.

The significant advantages of printed circuits are somewhat offset by the electromagnetic complexity of the structure, because its inherent inhomogeneity makes accurate calculations difficult. Microstrips have received much attention ever since their introduction in 1952. Many models were proposed for their analysis, from simple static and quasistatic approximations all the way to full-wave integral formulations solved by sophisticated computer

v

techniques. Accuracies reaching fractions of a percent can now be achieved, at the cost of computer time and memory.

Research in this field actively continues, because many effects are not yet satisfactorily characterized. In particular, interactions between components and lines require a detailed analysis of radiated and surface waves, which also interact with the sides and covers of the enclosures. After more than 40 years of activity, microstrip development is alive and well. With the increasing integration of circuits in MMICs (monolithic microwave integrated circuits) and the development of higher frequency bands in the millimeter-wave range, it more and more appears to be a never-ending process.

Over the last 40 years, numerous publications have been devoted to microstrips, and the flow is by no means ebbing. For many years, complete sessions of most microwave conferences and symposia have been devoted to the presentation of the latest results obtained for microstrip circuits and antennas. A good number of books, textbooks, and collections of articles were devoted to them already.

Several times, while presenting lectures on microstrips in various parts of the world, attendees asked me the following question: We are just starting a microstrip lab here, could you tell me what we should do first? This book is an attempt to answer this question, by providing a solid introduction and a general overview from basic concepts to more sophisticated recent developments. After gaining sufficient background information the newcomer should be able to build upon it and assimilate specialized technical literature and papers presented at symposia. Many facets of microstrips are presented in the book, including basic theory, design, fabrication, and measurement techniques. The text is completed by an extended bibliography, and some problems are proposed to the reader.

GENERAL ORGANIZATION OF THE BOOK

The introduction briefly describes the microstrip structures and provides a simple description of main principles. Both the advantages and the drawbacks of microstrip techniques are outlined in some detail.

The basic foundations of electromagnetics are presented in the second chapter, in which Maxwell's equations are introduced, followed by the boundary conditions that must be satisfied between different media, and by the electromagnetic properties of the materials that make up the structures. The links between time-domain and frequency-domain approaches are outlined.

Transmission line theory considers infinitely long straight structures in Chapter 3. The basic line parameters such as the effective permittivity and the characteristic impedance are derived. The simplified quasi-TEM approach is given first, followed by more complex developments that take into account dispersion and losses.

A whole chapter is devoted to the characterization of devices by means of matrices and equivalent circuits. Their basic properties are outlined, and the particular case of a section of line is considered. The matching process is briefly described, and the chapter ends with deembedding, a technique that separates the circuit parameters from those of its surroundings.

Actual transmission lines are neither straight nor uniform, and the discontinuities and end effects are considered in Chapter 5. Many models were used by different researchers, and an overview of the more significant results is provided.

Moving to more complex structures, Chapter 6 deals with junctions and couplers. The description of three-port and four-port devices is made in terms of the scattering matrix (*S*-parameters).

The seventh chapter considers resonant devices, and their use to discriminate signals with different kinds of filters.

Some background on transistor amplifiers and their design is given in Chapter 8. The principle of amplification is briefly reviewed, and the main properties of transistors of different types are listed. Stability, reflection, and noise effects are discussed.

The ninth chapter is devoted to a short presentation of semiconductor diodes and their use as signal detectors, mixers, control devices, and sources.

Microstrip lines are inherently open structures in which the electromagnetic field may extend (in principle) all the way to infinity. Adjacent structures interact, and signal leakage may produce spurious cross-coupling and resonances. These effects prove troublesome in many practical situations, so ways to predict them, prevent their occurrence, and compensate for them are required. A short introduction to the most sophisticated computer techniques for the analysis of microstrips is given in Chapter 10.

Radiation is unwanted in circuits but is used to advantage to realize microstrip antennas, briefly described in Chapter 11. It is shown that antennas and lines are basically incompatible, so the two functions should be clearly separated to obtain favorable operating conditions.

Microstrips are used to realize rather complex assemblies of components, for which experimental design becomes long, tedious, and inefficient. Computer-aided design packages allow designers to considerably reduce the time and effort required to obtain the desired circuit response. The principles of CAD are outlined in Chapter 12, describing some models used and the techniques of synthesis and optimization that permit one to draw the layout of the desired circuit.

The final product is of course always the microstrip circuit itself, which must now be fabricated physically. Chapter 13 lists the steps of the photolithographic process and the techniques used to insert and bond components within the circuit.

Measurements finally show whether the theories and the models used together with the fabrication process are accurate and dependable. Microstrips present a particular challenge in that the outputs of laboratory generators and test equipment are in coaxial line or waveguide, so transitions and connectors to microstrips are required. The circuit itself is seen through connectors and transitions that may introduce significant errors. The actual parameters of the circuit can now be determined accurately, since the effects of connectors are suppressed by calibration of the network analyzer. In Chapter 14, techniques to measure substrate parameters are also reviewed, and a section is devoted to the visualization of the fields in the close vicinity of circuits.

Problems are proposed to the readers in Chapters 1–9, and complete solutions are given toward the end of the book. An extended bibliography is provided.

ACKNOWLEDGMENTS

The information in this book was collected over many years through the research and teaching activities at the Laboratory of Electromagnetism and Acoustics (LEMA) of the École Polytechnique Fédérale of Lausanne, Switzerland. I wish here to thank the collaborators and students who participated in our activities. Particularly worth mentioning are the important contributions of Mr. Jean-François Zürcher, who is mostly active in circuit layout, antenna design, and near-field display, and of Professor Juan R. Mosig in the field of integral equations. I also wish to acknowledge the contributions of Dr. Anja Skrivervik, Dr. Munikoti Ramachandraiah, Professor Thomas Sphicopoulos, Dr. Richard Hall, Dr. Syed Bokhari, Dr. Miguel Keer, and Dr. Lionel Barlatey. I am particularly grateful to Dr. and Mrs. Hugh K. Smith, who carefully revised my manuscript and provided most valuable comments (any mistakes that may be left remain of course my own). A special word of thanks is due to Professor K. C. Gupta, who, during the three months he spent in Lausanne as visiting professor in 1976 (on leave from IIT Kanpur, India), got us really involved in microstrip design and put us on the right track.

Activities in microstrip circuits at LEMA started in 1971 under the sponsorship of the Hasler Foundation. Over the years, they were financially supported by the Swiss National Foundation, the Swiss Commission for the Encouragement of Scientific Research, the Swiss Office for Education and Science (Actions COST-213 and COST-223), and the European Space Agency.

Part of this material was gathered for intensive courses organized in 1987 (Lausanne) and 1988 (Montreux) by the company High Tech Tournesol. Some of it was also presented at a summer school in Saint-Jean de Maurienne, France, in 1989 and at a lecture series at the National Defense Academy in Yokosuka, Japan, also in 1989. This topic was also presented in many places as a lecture of the "Speaker's Bureau" of the Microwave Theory and Techniques (MTT) Society of the IEEE in 1988 and 1989.

A sabbatical leave proved most favorable to complete my text in quiet and congenial surroundings. I wish here to thank the authorities and colleagues of four institutions that invited me during that period: Centre National de Recherches en Telecommunications (CNET), La Turbie, France; Laboratory for Infrared and Microwave Devices (LDIM), Université Pierre et Marie Curie (Paris VI Jussieu); Laboratory of Signals and Systems, Ecole Superieure d'Electricité (Supelec), Gif-sur-Yvette, France; and International Center for Theoretical Physics (ICTP), Trieste, Italy.

Lastly, I am particularly grateful to the direction of the Ecole Polytechnique Federale, Lausanne, for the continuous support granted over many years to the Laboratory of Electromagnetism and Acoustics and its microstrip activities.

Pully, Switzerland   FRED GARDIOL

Contents

Nomenclature

NOTATION

Note: [1] means that a quantity is dimensionless

$\underline{\mathbf{A}}$	magnetic phasor–vector potential	[V-s/m]
A	point of Smith chart	
$\underline{A}$	scattering matrix term	[1]
$\underline{A}$	term of the chain matrix	[1]
a	array spacings in the x direction	[m]
a	resonator dimension	[m]
a	resonator radius	[m]
a_{eff}	effective resonator radius	[m]
$\underline{a}_i$	complex normalized wave	[W$^{1/2}$]
B	frequency bandwidth	[Hz]
B	parallel susceptance	[S]
$\mathbf{B}$	induction field	[V-s/m^2]
$\underline{\mathbf{B}}$	induction phasor–vector	[V-s/m^2]
B	point on Smith chart	
$\underline{B}$	scattering matrix term	[1]
$\underline{B}$	term of the chain matrix	[Ω]
b	array spacings in the y direction	[m]
b	resonator dimension	[m]
b	thickness of upper conductor	[m]
$\underline{b}_i$	complex normalized wave	[W$^{1/2}$]
C	capacitance	[F]
C	contour of integration	[m]
$\underline{C}$	term of the chain matrix	[S]
c	wave velocity	[m/s]
C'	capacitance per unit length	[F/m]

$C(U)$	junction capacitance	[F]
C_0	capacitance at zero bias	[F]
c_0	velocity of light $(2.997925\ldots \times 10^8)$	[m/s]
C_{p}	package capacitance	[F]
$\underline{C}_{\mathrm{stg}}$	generator stability circle center	
$\underline{C}_{\mathrm{stL}}$	load stability circle center	
D	directivity (antenna)	[1]
D	largest antenna dimension	[m]
$\mathbf{D}$	displacement field	[A-s/m^2]
$\underline{\mathbf{D}}$	displacement phasor–vector	[A-s/m^2]
$\underline{D}$	term of the chain matrix	[1]
d	corner width	[m]
dA	element of surface	[m^2]
dB	decibel	[1]
d**l**	element of contour	[m]
dV	element of volume	[m^3]
dz	infinitesimal section of line	[m]
e	base of neperian logarithm $(\cong 2.718)$	[1]
$\mathbf{E}$	electric field	[V/m]
$\underline{\mathbf{E}}$	electric phasor–vector	[V/m]
$\underline{\mathbf{E}}_+$	electric phasor–vector of the forward wave	[V/m]
$\underline{\mathbf{E}}_-$	electric phasor–vector of the backward wave	[V/m]
$\underline{\mathbf{E}}_0$	electric field in cavity	[V/m]
$\mathbf{E}^{\mathrm{e}}$	excitation field	[V/m]
$\mathbf{E}^{\mathrm{s}}$	scattered field	[V/m]
E_u	electric field component along coordinate u	[V/m]
E_{u0}	effective amplitude of u component	[V/m]
E_{UV}	error term (calibration)	[1]
$\underline{F}(\omega)$	Fourier transform (spectrum)	
(F)	diagonal auxiliary matrix	[S$^{-1/2}$]
F	noise figure	[1]
$F_{\min}$	minimum noise figure	[1]
f	signal frequency	[s^{-1}]
f_c	cutoff frequency (higher-order mode)	[s^{-1}]
$f_c(U)$	cutoff frequency (varactor)	[s^{-1}]
f_r	resonant frequency (cavity)	[s^{-1}]
f_{mn}	resonant frequency	[s^{-1}]
F(Φ)	variational principle	
$\mathbf{F}_{\mathrm{e}}$	electric force	[N]
$\mathbf{F}_{\mathrm{m}}$	magnetic force	[N]
$\underline{G}(\omega)$	response signal (Fourier transform)	
$\underline{\underline{G}}_V(\mathbf{r}|\mathbf{r}')$	Green's function for scalar potential	[V/A-s]
$\overline{\mathbf{G}}_{\mathrm{A}}(\mathbf{r}|\mathbf{r}')$	Green's function for vector potential	[V-s/A-m^2]
G_1	gain (antenna)	[1]
G_A	free-space Green's function (vector potential)	[V-s/A-m^2]

G_T	transducer gain	[1]
G_{Tu}	unidirectional transducer gain	[1]
(G)	diagonal impedance matrix	[Ω]
$\underline{H}(\omega)$	transfer function	
H	magnetic field	[A/m]
$\underline{\mathbf{H}}$	magnetic phasor–vector	[A/m]
$h(t)$	impulse response	
h	substrate thickness	[m]
h	Planck's constant (6.62×10^{-34})	[J-s]
$\underline{I}(z)$	transmission line current	[A]
I_0	dc component of current (rectifier)	[A]
$\underline{I}_i$	current at port i	[A]
I_s	saturated inverse current (diode)	[A]
i_{in}	input current	[A]
i_{out}	output current	[A]
$\underline{\mathbf{J}}$	current phasor–vector	[A/m^2]
J	electric current density	[A/m^2]
$\mathbf{J}_s$	surface current density	[A/m]
k_B	Boltzmann constant (1.3804×10^{-23})	[J/K]
L	inductance	[H]
L	length of resonator	[m]
L	length of transmission line	[m]
L	coefficient (finite elements)	[V/m]
L'	inductance per unit length	[H/m]
LA	insertion loss	[dB]
LC	coupling	[dB]
LD	directivity	[dB]
LF	noise figure	[dB]
LI	isolation	[dB]
L_g	effective gate length	[m]
$L\mathrm{R}$	reflected power	[dB]
L_s	wire inductance	[H]
M	coefficient (finite elements)	[V/m]
M	magnetization field	[A/m]
m	integer	[1]
m	mode index	[1]
N	average noise power	[W]
N	coefficient (finite elements)	[V]
n	mode index	[1]
n	number of device ports	[1]
n	vector perpendicular to an interface	[m]
Np	Nepers	[1]
$P(z)$	active power on transmission line	[W]
P	active power density	[W/m^2]
P	polarization field	[A-s/m^2]

P_b	bias power	[W]
P_{in}	incident power	[W]
P_{in}	input power	[W]
P_{out}	output power	[W]
P_r	received power (antenna)	[W]
P_r	reflected power	[W]
P_t	transmitted power (antenna)	[W]
Q	quality factor	[1]
$Q(z)$	reactive power on transmission line	[V-A]
Q_0	unloaded quality factor	[1]
$Q(U)$	quality factor (varactor)	[1]
$\mathbf{Q}$	reactive power density	[W/m^2]
q	charge of the electron	[A-s]
q	electric charge	[A-s]
R	distance between two antennas	[m]
R	median radius of ring resonator	[m]
r	spherical coordinate	[m]
r	observation point	[m]
$\mathbf{r}$	vector indicating position in space	[m]
$\mathbf{r}'$	coordinate of sources	[m]
$\mathbf{r}'$	source point	[m]
$\mathbf{r}_t$	transverse part of position vector	[m]
R_L	load resistance	[Ω]
R_m	metal resistance	[Ω]
R_n	noise resistance	[Ω]
R_s	series resistance	[Ω]
R_{stg}	generator stability circle radius	
R_{stL}	load stability circle radius	
$\underline{S}(z)$	complex power on transmission line	[V-A]
$\underline{\mathbf{S}}$	Poynting phasor–vector	[W/m^2]
S_{PQR}	triangle area (finite elements)	[m^3]
S	surface of integration	[m^2]
s	chamfer	[m]
$\underline{s}_{ij}$	scattering matrix parameter	[1]
T	equivalent network	
T	junction	
T	equivalent noise temperature	[K]
T	period of sinusoidal signal	[s]
T	physical junction temperature	[K]
$[\underline{T}^k]$	connecting chain matrix	
T_0	standard noise temperature (290)	[K]
t	time	[s]
$\underline{U}(z)$	transmission line voltage	[V]
$\underline{U}_i$	voltage at port i	[V]
U	voltage across the junction	[V]

V	cavity volume	$[m^3]$
V	volume of integration	$[m^3]$
$\underline{V}$	scalar potential (phasor)	$[V]$
V'	volume of sources	$[m^3]$
$\mathbf{v}$	velocity	$[m/s]$
v_ϕ	phase velocity	$[m/s]$
v_s	saturation velocity	$[m/s]$
VSWR	voltage standing-wave ratio	$[1]$
w	microstrip width	$[m]$
$\underline{w}$	conformal mapping function	
w_{eff}	effective width	$[m]$
w_b	broad-strip width of low-pass filter	$[m]$
w_e	effective width of microstrip	$[m]$
w_n	narrow strip of low-pass filter	$[m]$
x	rectangular coordinate	$[m]$
x_0	coordinate of reference element (array)	$[m]$
X_L	load reactance	$[\Omega]$
Y	junction	
y	rectangular coordinate	$[m]$
y_0	coordinate of reference element (array)	$[m]$
$\underline{Y}_c$	characteristic admittance	$[S]$
$\underline{Y}_e$	even-mode admittance	$[S]$
$\underline{Y}_{ij}$	term of the admittance matrix	$[S]$
$\underline{Y}_{min}$	admittance for minimum noise figure	$[S]$
$\underline{Y}_o$	odd-mode admittance	$[S]$
$\underline{Y}_s$	source admittance	$[S]$
$\underline{Y}_{aa}$	antisymmetrical–antisymmetrical admittance	$[S]$
$\underline{Y}_{as}$	antisymmetrical–symmetrical admittance	$[S]$
$\underline{Y}_{sa}$	symmetrical–antisymmetrical admittance	$[S]$
$\underline{Y}_{ss}$	symmetrical–symmetrical admittance	$[S]$
$\underline{z}$	conformal mapping function	
z	rectangular coordinate	$[m]$
z_i	coordinate axis at port i	$[m]$
Z_0	free-space impedance (120π)	$[\Omega]$
Z_c	characteristic impedance	$[\Omega]$
Z_{ci}	characteristic impedance at port i	$[\Omega]$
$\underline{Z}_e$	even-mode impedance	$[\Omega]$
$\underline{Z}_{ij}$	term of the impedance matrix	$[\Omega]$
$\underline{Z}_{in}$	input impedance	$[\Omega]$
$\underline{Z}_L$	load impedance	$[\Omega]$
$\underline{Z}_m$	metal impedance	$[\Omega]$
$\underline{Z}_o$	odd-mode impedance	$[\Omega]$
$[\underline{Z}]$	intrinsic impedance matrix	$[\Omega]$
$[\underline{Z}^k]$	connecting impedance matrix	$[\Omega]$
$[Z_{tot}]$	global impedance matrix	$[\Omega]$
Z_{out}	output impedance	$[\Omega]$
Z_T	transformer impedance	$[\Omega]$

MATHEMATICAL SYMBOLS

$\dfrac{\partial}{\partial}$	partial derivative	
$\partial/\partial t$	time derivative	$[\text{s}^{-1}]$
∇	differential operator (del or nabla)	$[\text{m}^{-1}]$
$\nabla \times$	**curl**	$[\text{m}^{-1}]$
$\nabla \cdot$	divergence	$[\text{m}^{-1}]$
∇_t	transverse differential operator	$[\text{m}^{-1}]$
$\mathbf{1}(t)$	Heaviside unit step	
$*$	convolution product	$[1]$
$\times$	vector product (cross product)	$[1]$
$\cdot$	scalar product (dot product)	$[1]$

GREEK SYMBOLS

α	attenuation per unit length	$[\text{m}^{-1}]$
α	insertion loss term (coupler)	$[1]$
α	voltage factor	$[\text{V}^{-1}]$
α_d	dielectric attenuation per unit length	$[\text{m}^{-1}]$
α_m	conductor attenuation per unit length	$[\text{m}^{-1}]$
$\underline{\alpha}$	scattering matrix term	$[1]$
β	coupling term	$[1]$
β	propagation factor	$[\text{m}^{-1}]$
β_e	propagation factor of even mode	$[\text{m}^{-1}]$
β_o	propagation factor of odd mode	$[\text{m}^{-1}]$
$\underline{\beta}$	scattering matrix term	$[1]$
γ	capacitance exponent (diode)	$[1]$
γ	propagation factor (phasor–vector)	$[\text{m}^{-1}]$
$\underline{\gamma}$	scattering matrix term	$[1]$
$\underline{\Gamma}_\text{g}$	reflection factor of generator	$[1]$
$\underline{\Gamma}_\text{L}$	reflection factor of load	$[1]$
δ	infinitesimally small dimension (boundary)	$[\text{m}]$
δ	skin depth	$[\text{m}]$
$\delta(t)$	Dirac pulse	
Δl	equivalent line length	$[\text{m}]$
Δu	finite difference	
ΔV	sample volume	$[\text{m}^3]$
$\Delta \omega$	intermediate angular frequency	$[\text{s}^{-1}]$
Δz_i	shift of reference plane	$[\text{m}]$
ε	absolute permittivity	$[\text{A-s/V-m}]$
$\underline{\varepsilon}$	complex permittivity	$[\text{A-s/V-m}]$
$-\varepsilon''$	imaginary part of complex permittivity	$[\text{A-s/V-m}]$
ε'	real part of complex permittivity	$[\text{A-s/V-m}]$
$\varepsilon(t)$	permittivity impulse response	$[\text{A-s/V-m}]$
ε_0	electric constant of vacuum	$[\text{A-s/V-m}]$

ε_e	effective permittivity	[1]
$\varepsilon_{ef}(f)$	frequency-dependent effective permittivity	[1]
ε_{es}	low-frequency effective permittivity	[1]
ε_r	relative permittivity	[1]
$\overline{\overline{\varepsilon}}$	permittivity tensor	[A-s/V-m]
η	efficiency (antenna)	[1]
η	ideality factor (diode)	[1]
η_p	partial efficiency	[1]
η_{pa}	added power efficiency	[1]
η_t	total efficiency	[1]
$\underline{\Theta}$	transverse magnetic potential	[A]
$\underline{\theta}$	phase	[1]
θ	spherical coordinate	[1]
κ	conductor loss factor	[$s^{1/2}$/m]
λ_0	free-space wavelength	[m]
λ_g	line wavelength	[m]
μ	absolute permeability	[V-s/A-m]
$\underline{\mu}$	complex permeability	[V-s/A-m]
$-\mu''$	imaginary part of complex permeability	[V-s/A-m]
μ'	real part of complex permeability	[V-s/A-m]
$\mu(t)$	permeability impulse response	[V-s/A-m]
μ_0	magnetic constant of vacuum	[V-s/A-m]
$\underline{\mu}_r$	relative permeability	[1]
$\overline{\overline{\mu}}$	permeability tensor	[V-s/A-m]
Π	equivalent network	
$\underline{\rho}$	charge phasor	[A-s/m^3]
$\overline{\rho}$	electric charge density	[A-s/m^3]
ρ_s	surface charge density	[A-s/m^2]
$\underline{\rho}_{aa}$	antisymmetrical–antisymmetrical reflection factor	[1]
$\underline{\rho}_{as}$	antisymmetrical–symmetrical reflection factor	[1]
$\underline{\rho}_{sa}$	symmetrical–antisymmetrical reflection factor	[1]
$\underline{\rho}_{ss}$	symmetrical–symmetrical reflection factor	[1]
σ	conductivity	[S/m]
$\underline{\sigma}_{sc}$	complex conductivity (superconductor)	[S/m]
τ	time delay	[s]
τ	transit time	[s]
$\underline{\tau}$	scattering matrix term	[1]
Φ	barrier potential	[V]
$\underline{\Phi}$	transverse electrical potential	[V]
ϕ	spherical coordinate	[1]
ϕ_u	phase of u component of the electric field	[1]
χ_{mn}	mth zero of the Bessel function J_n	[1]
ψ	phase	[1]
Ψ_{mn}	phase shift between array elements	[1]
ω	angular frequency ($2\pi f$)	[s^{-1}]
ω_{if}	intermediate angular frequency	[s^{-1}]
ω_{lo}	local oscillator angular frequency	[s^{-1}]

CHAPTER ONE

Introduction

Microstrips are printed circuits for very high frequency electronics and microwaves. When made of conducting strips deposited upon a dielectric substrate, they are called microwave integrated circuits (MICs). Printed circuits are also made with semiconducting substrates, inside of which electronic devices (transistors, diodes) are fabricated by diffusion or ion implantation. The resulting structures are then called monolithic microwave integrated circuits (MMICs).

1.1 MICROSTRIP STRUCTURES

1.1.1 Basic Description

Physically, any microstrip structure consists of a thin plate of low-loss insulating material, the *substrate*, covered with metal completely on one side, the *ground plane*, and partly on the other, where the circuit or antenna patterns are printed (Figure 1.1). Lumped components can be connected into the circuits (discrete components, Section 13.5) or realized within the circuit (distributed components, Section 13.9).

1.1.2 The Substrate

The substrate fulfills two functions:

1. It is a mechanical support that ensures that implanted components are properly positioned and mechanically stable, just as in printed circuits for low-frequency electronics.
2. It behaves as an integral part of connecting transmission lines and deposited circuit components: its permittivity and thickness determine the electrical characteristics of the circuit or antenna.

1

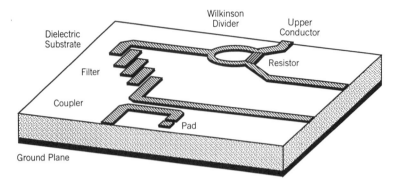

FIGURE 1.1 Microstrip structure, showing the conductor pattern and various inserted circuit components.

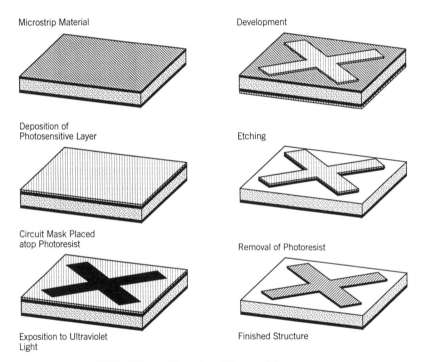

FIGURE 1.2 The photolithographic process.

1.1.3 Metallization

Most commercially available microstrip substrates are metallized on both faces. The circuit pattern is realized by the photolithographic process (Chapter 13). A mask of the circuit to be realized is drawn at a suitable scale, cut, and then reduced and placed on top of a photoresistive layer, which was previ-

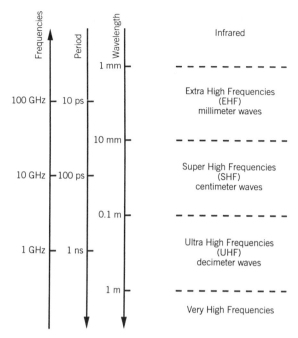

FIGURE 1.3   The microwave range of frequencies.

ously deposited on top of the microstrip. The structure is then exposed to ultraviolet radiation, which reaches the photosensitive layers through the mask openings. The exposed parts are removed by the photographic development, and the metal cover is etched away from the exposed area (Figure 1.2). It is also possible to deposit metal by evaporation or sputtering upon a bare dielectric substrate.

1.2 MICROWAVES

1.2.1 Frequency Range

The frequency ranges extending from 300 MHz to 300 GHz, characterizing signals that vary periodically between 300 million and 300 billion times per second (3×10^8–3×10^{11} Hz), are called *microwaves*. These limits are somewhat arbitrary, indicating a general location "above" the ranges used for radio and television broadcasting and "below" the infrared and visible rays (Figure 1.3). By extension, the term *microwaves* is used to characterize equipment and particular techniques developed for the exploitation of that specific frequency range.

1.2.2   Period and Wavelength

The *period* T of a microwave signal, defined by the inverse of its frequency f, is between about 3 nanoseconds (ns) and 3 picoseconds (ps). For a plane wave propagating in free space at the velocity of light ($c_0 = 2.997925\ldots 10^8$ m/s), the wavelength λ_0, defined by $\lambda_0 = c_0\,T = c/f$, is between a millimeter (mm) and a meter (m).

1.2.3   Remark about Dimensions

At microwaves, the technical equipment used to generate and process signals has a size similar to the wavelength. Components cannot be assumed to be dimensionless points in space, with signals traveling instantaneously, as is done at low frequencies (circuit theory).

Further up the frequency scale is the optical range, in which devices become very much larger than a wavelength. Other approximations can then be made, leading to geometric optics and ray techniques.

These approximations are not valid in the microwave range, and problems must be treated in terms of electric and magnetic fields. Maxwell's equations must be solved in the presence of boundary and initial conditions (Chapter 2). In particular, the voltage cannot be defined uniquely, because the electric field does not derive from a potential. Signals having a sinusoidal time dependence are considered, because use of the complex notation leads to important simplifications (phasor vectors, Section 2.4.2).

A similar situation is encountered in acoustics. The velocity of sound in air is about 300 m/s: acoustic waves travel roughly a million times slower than light. Wavelengths between 1 mm and 1 m correspond to sound frequencies between 300 Hz and 300 kHz, over which most acoustic sources operate: human voice, music. Since wavelengths of the same order of magnitude are considered in the two fields, microwaves and acoustics present dimensional similarities, even though the physical mechanisms involved are different. Techniques developed in one field can be transposed into the other.

1.2.4   Microwave Properties

Within the microwave range, nature provides particularly favorable conditions for signal propagation:

The ionosphere, made of electron sheaths surrounding the earth, is transparent to microwave signals, whereas lower-frequency signals get reflected.

The atmosphere is also transparent for microwaves, except over the higher-frequency ranges (above 10 GHz) where atmospheric constituents (rain, fog, snow) contribute absorption and diffraction that may become severe in bad weather.

The electromagnetic noise power picked up by an antenna pointed skyward reaches a minimal value within the microwave range. Receivers exhibit there the largest possible sensitivity, allowing one to detect extremely weak signals, for instance those coming from deep-space probes (on August 25, 1989, *Voyager 2* passed by the planet Neptune, at about 4.5 billion km from earth and sent colored pictures of the planet and of its satellite Triton).

Since the directivity depends on the ratio of size to wavelength, microwave antennas are directive without becoming prohibitively large.

The effective reflection from a target also depends in a sensitive manner on the size-to-wavelength ratio. Microwaves are thus well suited to detect obstacles having sizes on the order of meters (planes, ships, cars, people) but are not affected by raindrops.

Microwaves are absorbed by water and thus can be used to dry, heat, process materials, and cook food. Their absorption can also be used to determine the amount of moisture contained within materials.

The energy of microwave photons lies within the range 1 μeV to 1 meV, that is, several orders of magnitude below the energy of a molecular bond. This means that microwaves are nonionizing (one cannot get suntanned with microwaves).

The most stable natural oscillators known are located within the microwave range: hydrogen, cesium, rubidium atoms. Most frequency standards and atomic clocks operate, at least partly, at microwave frequencies (hydrogen maser, cesium beam, rubidium cell).

1.2.5 Application of Microwaves

In many applications, microwaves are essential because their specific properties are an absolute must. The microwave field was developed during World War II for radars, which require high-directivity antennas and adequate reflections from obstacles in the meter range.

Later, their use was extended to communications, both at the ground level (microwave links) and in space, following the successful launch of orbiting satellites. Radars and communications operate mainly over the frequency ranges below 15 GHz, with some applications at frequencies up to 100 GHz.

Microwave heating became a significant application in the 1960s, with the increasing acceptance by the general public of the domestic microwave oven (which operates in the vicinity of 2450 MHz). Microwaves are also used for medical and industrial heating at 2450 MHz and, in some countries, at 915 MHz (ISM bands). Their interactions with materials are used for measurements: microwave spectroscopy, moisture content.

Computer systems operate with increasingly faster switching devices, with data rates reaching up to several gigabits per second. Logic frequency dividers were reported for operation up to 18 GHz. Digital electronics thus require the use of microwave techniques [Wilhelm 1986]. When the transit time along a connecting strip is similar to a period or to the duration of a pulse, circuits must be analyzed in terms of transmission lines, characterized by their geometrical outline. Discontinuities and mismatches may then produce multiple reflections and thus create complex echoes whose effects tend to be detrimental for proper system operation [Gardiol 1987].

1.2.6 Metal Waveguides

Microwave signals were traditionally transmitted through hollow metal pipes called waveguides. Waves bounce back and forth on the conductor walls, producing signal propagation when frequencies exceed a lower bound called the cutoff frequency [Montgomery et al. 1948]. Metal waveguides are inherently rigid; that is, the realization of waveguide circuits leads to some interesting topological problems. The circuits themselves tend to be rather heavy and bulky. Waveguides are therefore used mainly where high-power operation is required (transmitters, heating applications) and at the upper limits of the microwave range (millimeter waves).

1.2.7 Printed Circuits

Whenever the frequency range or the power level of the signal permits, waveguides are replaced by printed circuits (Figure 1.4). Open structures are generally used for the lower-frequency ranges (microstrip, slotline, coplanar

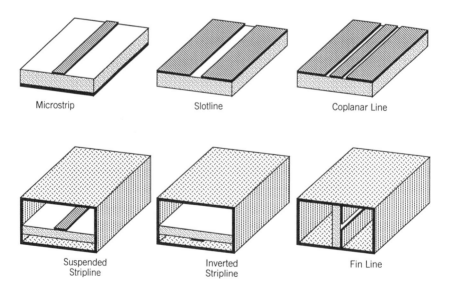

FIGURE 1.4 Printed circuits for microwaves.

line), whereas toward higher ranges the circuits must be surrounded by a metallic enclosure to prevent radiation (printed circuit within a waveguide: suspended line, inverted line, finline). Microstrips, first described in 1952 in two companion articles [Grieg and Engelmann 1952; Assadourian and Rimai 1952], have become the structures most commonly used to realize MICs and MMICs.

1.3 WAVES IN MICROSTRIPS

1.3.1 Point Source on a Microstrip

The physical features producing transmission and radiation in planar lines can be understood by considering a point current source (Hertz dipole) located on the edge of the upper conductor (Figure 1.5). This source radiates electromagnetic waves in all directions. Depending on the angles toward which they are transmitted, the waves fall within four distinct categories and exhibit widely different behaviors.

1.3.2 Guided Waves

For an angle located between "six and nine o'clock" (defining directions as in a watch), waves are guided within the substrate by reflections on the two metal walls. They propagate in a parallel-plate waveguide and provide the normal operation of transmission lines and circuits (Chapters 3, 4, and 5). On the other hand, these waves contribute an unwanted buildup of electromag-

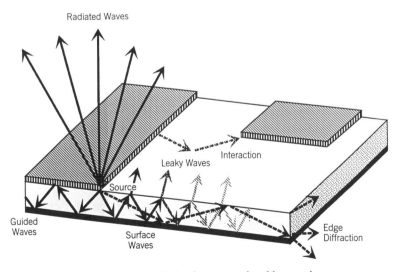

FIGURE 1.5   The four kinds of waves emitted by a point source.

netic energy under antenna patches that act as resonators with a limited bandwidth.

When the substrate is thin (as compared with a wavelength) and has a large permittivity, the guided waves are predominant.

1.3.3 Radiated Waves

Waves directed toward angles between "nine and three o'clock" go upward, directly toward the open space above, where they do not meet any further boundaries. They contribute radiation and are thus required for antenna operation (Chapter 11). In transmission lines and circuits they produce leakage and are therefore unwanted.

Radiated waves are predominant when the substrate is thick (as compared with a wavelength) and has a low permittivity.

1.3.4 Leaky Waves

Within the "three to six o'clock" quadrant, the ground plane reflects waves toward the air–dielectric interface. Depending on their angle of incidence, some waves are partly reflected and partly transmitted (leaky waves), while others are totally reflected (surface waves). The leaky waves contribute to radiation and may, under favorable conditions, increase the apparent antenna size and thus produce a larger directivity [Jackson and Oliner 1988].

1.3.5 Surface Waves

With a larger incidence angle, the surface waves remain trapped by total reflection within the dielectric substrate, where they propagate with small loss. Their effects may be troublesome for lines and circuits as well as for antennas. The signal amplitude is reduced, creating an apparent attenuation or a decrease in antenna gain and efficiency [Bhattacharyya and Shafai 1986].

Surface waves also produce spurious coupling between circuit or antenna elements. This feature is particularly obnoxious in phased array antennas, which can neither transmit nor receive in some directions (blind spots) [Pozar and Schaubert 1984]. When surface waves reach the far edges of the structure, they are diffracted and contribute spurious radiation, degrading the antenna pattern (side lobes, cross-polarization). Surface waves become significant when the substrate is thick (as compared with a wavelength) and has a large permittivity.

1.3.6 Circuit and Antenna Requirements

Depending on the relative amplitude of the excited waves, a structure behaves like a transmission line, an antenna, or a launcher of surface waves.

The electromagnetic field in a transmission line or in a circuit must remain in the close vicinity of the conductors. The guided waves are to be excited preferentially, while one should avoid exciting radiated, leaky, and surface waves as much as possible.

Quite the opposite is required in an antenna, where radiated waves are preferred and where the geometry is adjusted to prevent energy concentration by guided waves below the patch. At the same time, surface waves must be avoided.

The substrate requirements for circuit operation are inconsistent with those needed for antenna operation. It is not possible to realize an efficient antenna and a nonradiating circuit upon the same substrate.

Rays were used in Figure 1.4 to depict the four kind of waves. Their use is valid when the wavelengths are much smaller than the dimensions of the structure, and this is not the case for printed circuits or antennas. A rigorous study must consider Maxwell's equations (Chapter 2) or the wave equations that are derived from them. The radiation from microstrips is considered in Chapters 10 and 11.

1.4 SIGNIFICANT ADVANTAGES

1.4.1 Reduction in Size

The use of printed circuits reduces the size and normalizes the designs, as compared with the more cumbersome waveguide parts. The size reduction results from the substrate permittivity: size is reduced roughly by the square root of the effective permittivity (Chapter 3). The resulting circuits are rugged and lightweight. Printed circuits, based on microstrip, slot, and coplanar lines, have replaced most waveguide assemblies for low power levels at frequencies below 12 GHz. The availability of low-loss dielectrics now allows one to use them up to 25 GHz, with future extensions into the millimeter-wave range [Williams 1991]. At lower-frequency ranges (200–400 MHz), the use of substrates having very high relative permittivities (up to $\varepsilon_r = 80$) leads to considerable size reductions for printed circuit elements [Tamura et al. 1988].

1.4.2 Large-Scale Fabrication

Waveguide designs call for machining of intricate parts, which in most situations must be fabricated one at a time and then assembled, brazed, or soldered. In contrast, complete printed circuit assemblies can be produced rather easily in large quantities, once the basic circuit pattern has been developed and tested successfully.

1.4.3 Insertion of Components

Active or passive lumped-circuit elements, dielectric resonators, and even antennas can be easily inserted (soldered, glued) or combined with microstrip circuits, taking advantage of the fact that the structure is open (Chapter 13). In rather sharp contrast, the introduction of any device within a waveguide is a major operation, requiring the machining of suitable supports or adapters.

1.4.4 Monolithic Integrated Circuits

Printed circuit techniques have in turn led to the development of integrated circuits, in which metal strips are deposited directly on top of the semiconductor material in which active functions (diodes, transistors, gates, lasers) are carried out. Monolithic integrated circuits are also developed at microwaves, presenting particular challenges due to their small dimensions and to the high-density packing of lines and components [Pucel 1986].

1.5 MAIN DESIGN PROBLEMS

1.5.1 Accuracy Requirements

At low frequencies, the metal strips must only provide a good galvanic connection with low resistance, the dimensions being relatively uncritical. At microwave frequencies, on the other hand, all dimensions become significant: the width of a strip is directly related to line impedance, and hence to match and reflections, while its length determines the phase shift. Care is required in the fabrication to ensure proper circuit operation.

1.5.2 Electromagnetic Complexity

Thanks to Maxwell, electromagnetism is an exact science. The rigorous analysis of microstrip structures requires the resolution of Maxwell's equations in the presence of boundary conditions (Section 2.2). This analysis becomes quite involved because for electromagnetic fields microstrip structures are three times inhomogeneous (Figure 1.6):

a. The electromagnetic fields extend over different propagation media: air and dielectric or air and several dielectrics.
b. The boundary conditions on the air–dielectric interface are transversely inhomogeneous, because this surface is partly covered with metal. Since the conducting layer is very thin, it is often replaced in the mathematical developments by an equivalent sheet of surface current.

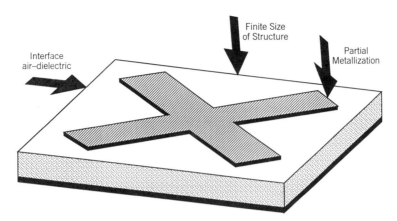

FIGURE 1.6 Microstrip circuit, schematically showing the three kinds of inhomogeneities.

c. The devices exhibit finite dimensions in the transverse plane. The transverse boundaries may be located far enough from the conducting strips, in which case edge effects may be neglected and the interface considered to be of infinite size.

Due to this threefold inhomogeneity, a rigorous analysis is difficult to carry out. Integral techniques coupled with a numerical resolution on a computer lead to results very close to measured values (Chapter 10).

1.5.3   Connector Mismatches

Printed circuits are assembled with coaxial transitions and connectors (Figure 1.7). Transitions are particular devices that connect different input and output lines. Whenever possible, one would like to achieve a smooth transition from input to output, but in many situations sharp discontinuities cannot be avoided. Matching is then required to reduce the reflections.

Whenever feasible, one should assemble all the components on a single substrate. It is then not possible to measure first the different parts of the circuit and connect them, as one would normally do with waveguide designs.

1.5.4   Transition Effects

Measurements are seldom made directly upon the microstrip structure itself because the equipment available connects to coaxial line or waveguide. This means that the microstrip circuit under test is practically always "seen" through connectors and transitions (Section 1.5.3).

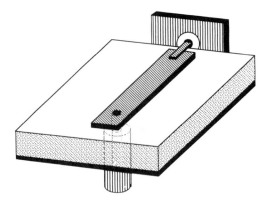

FIGURE 1.7 Microstrip to coaxial line transitions.

Transitions contribute reflections, attenuation, and phase shift that combine in a complex manner with the parameters of the circuit under test. The measurement data thus include the effect of the transitions, and one must determine what the characteristics of the circuit would be if there were no connectors or transitions. This process is called deembedding (Section 4.8). One may compare the circuit to a uniform transmission line of known length, assuming that the connectors used in both measurements are identical. One may also insert the element under test into a resonant ring and analyze the resonance pattern. The time-gating function of vector network analyzers allows one to discriminate between circuit and connector properties: by taking the inverse Fourier transform of a broad-band frequency response, one obtains a time-domain response in which contributions from different parts of the circuit can be separated. *The effect* of the connectors is removed, and a Fourier transform provides the frequency response without the effect of connectors. The effect of transitions can also be compensated for by calibration (Section 14.1.4). However, it is not possible to make accurate measurements across highly lossy or badly mismatched transitions.

1.5.5 Absence of Tuning Capability

Waveguide assemblies often require "last minute" adjustments: tuning screws are provided for this purpose, while inductive spring-loaded posts or dielectric bits and pieces can be inserted in the waveguide to compensate for a mismatch. Most waveguide filters must be individually tuned.

Similar tuning and fine adjustments are practically impossible on printed or integrated circuits: there is no such thing as a "tuning screw" for microstrip. The fabrication process is quite lengthy and involved (Chapter 13). If the completed circuit does not function as expected, the whole process must be repeated, after determining the most probable cause of the malfunction and hoping that it will work better the next time!

Particular precautions must therefore be taken to ensure that circuits function properly without need for individual adjustments. Their operation should not be significantly affected by small changes in component values, and the most suitable designs are selected by sensitivity analysis. To do this, one must be able to characterize the components accurately.

1.5.6 Effect of Enclosure

The theoretical derivations consider lines and circuits on an open infinite microstrip: strips are deposited on a substrate extending (theoretically) to infinity over the transverse x, y plane. Actual circuits are, however, enclosed in a metal box. The presence of a cover and side walls may significantly modify the performance, particularly when resonant modes of the box get excited (cavity effect). Placing the cover at least ten times the substrate thickness above the circuit is generally recommended [Laverghetta 1991].

1.5.7 Radiation and Surface Waves

In open microstrip circuits, radiation and surface waves are excited by currents flowing in the conductors (Section 1.3). When the circuits are enclosed in metal boxes, the waves bounce back and forth on the walls and may produce spurious interference between different parts of a circuit. These effects increase with frequency and become more significant as signals are transmitted at faster rates. To some extent, they can be avoided by carefully selecting the substrate, and may be reduced by adding metallic or absorbing barriers.

1.5.8 High-Frequency Problems

As the operating frequency of microstrip circuits increases, the thickness of the dielectric substrate must be reduced accordingly to prevent radiation and excitation of surface waves. However, substrates cannot be thinned indefinitely, so some amount of surface waves and radiation cannot be avoided when operating in the millimeter-wave region. The microstrip line and its enclosure must then be considered together, under the names of suspended or inverted lines (Figure 1.4). Since the circuits become rather small, one must reduce the permittivity of the substrate, for instance by using part dielectric and part air.

In the standard analysis of microstrip, it is generally assumed that the upper conductor is infinitely thin. This assumption no longer holds at millimeter waves, for instance when a 2-μm-thick metal strip is deposited atop a 5-μm-thick substrate. At the higher frequencies, the finite thickness of the strips must be taken into account.

1.6 PROBLEMS

1.6.1 Determine the period and the wavelength in free space of a signal having a frequency of 465 MHz.

1.6.2 What is the frequency of a signal that has a free-space wavelength of 22 cm (a) in microwaves? (b) in acoustics?

1.6.3 Determine, approximately, the reduction in size that would be obtained when a substrate having a relative permittivity of 2.3 is replaced with one having a relative permittivity of 80.

CHAPTER TWO

Basic Electromagnetics

The whole field of electromagnetics is built upon a group of linear equations that were set up by James Clerk Maxwell (1831–1879). Four equations provide a compact unified formulation for the interactions between electric and magnetic fields, currents and charges, that were previously discovered and described by Ampere, Gauss, Faraday, and many others, mostly in the first half of the nineteenth century.

2.1 MAXWELL'S EQUATIONS

2.1.1 Electric and Magnetic Fields

The electromagnetic interactions involve six quantities, called fields, which are uniquely defined everywhere in space:

$$\mathbf{E} = \text{electric field} \qquad [\text{V/m}],$$
$$\mathbf{D} = \text{displacement field} \qquad [\text{A-s/m}^2],$$
$$\mathbf{H} = \text{magnetic field} \qquad [\text{A/m}],$$
$$\mathbf{B} = \text{induction field} \qquad [\text{V-s/m}^2],$$
$$\mathbf{J} = \text{electric current density} \qquad [\text{A/m}^2],$$
$$\rho = \text{electric charge density} \qquad [\text{A-s/m}^3]. \qquad (2.1)$$

The first five of these six quantities are vectors (in bold type), represented by arrows in a three-dimensional space, and defined either by three components in a suitable coordinate system, or by a direction and an amplitude (Figure 2.1). The last one is a scalar quantity, defined by a single term.

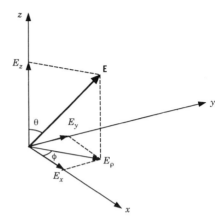

FIGURE 2.1 Definition of a vectorial field.

When applied to an electric charge q, the electric field produces an electric force $\mathbf{F_e}$:

$$\mathbf{F_e} = q\mathbf{E}. \tag{2.2}$$

This force can be measured, and when the value of the charge q is known, the amplitude and the direction of the electric field can be determined.

In addition, when the electric charge q moves with a velocity $\mathbf{v}$, the induction field $\mathbf{B}$ produces a magnetic force $\mathbf{F_m}$:

$$\mathbf{F_m} = q\mathbf{v} \times \mathbf{B}. \tag{2.3}$$

This force can also be measured, and when the charge q and the velocity $\mathbf{v}$ are known, the amplitude and the direction of the induction field $\mathbf{B}$ can be determined.

2.1.2 Local Form

In their local, or differential form, Maxwell's equations state that, at any point in space *that is not located on a boundary*, the electric and the magnetic field quantities defined in (2.1) must satisfy a set of four equations:

$$\nabla \times \mathbf{E} = -\frac{\partial \mathbf{B}}{\partial t}, \qquad \nabla \cdot \mathbf{D} = \rho,$$

$$\nabla \times \mathbf{H} = \frac{\partial \mathbf{D}}{\partial t} + \mathbf{J}, \qquad \nabla \cdot \mathbf{B} = 0. \tag{2.4}$$

This is the "local," or differential, form of Maxwell's equations, that links

space and time derivatives of the fields with electric current and charge·
densities. The differential operator ∇ (del or nabla) is used to define the two
differential operations $\nabla \times$ = **curl** and $\nabla \cdot$ = div.

2.1.3   Global Form

Maxwell's equations may also be expressed by their global or integral form, in
which four integral relationships involve field quantities over volumes, surfaces, and line contours:

$$\oint_C \mathbf{E} \cdot d\mathbf{l} = -\int_S n \cdot \frac{\partial \mathbf{B}}{\partial t} \, dA \quad [\text{V}],$$

$$\oint_C \mathbf{H} \cdot d\mathbf{l} = \int_S \mathbf{n} \cdot \left(\frac{\partial \mathbf{D}}{\partial t} + \mathbf{J} \right) dA \quad [\text{A}],$$

$$\oint_S \mathbf{n} \cdot \mathbf{D} \, dA = \int_V \rho \, dV \quad [\text{A-s}],$$

$$\oint_S \mathbf{n} \cdot \mathbf{B} \, dA = 0 \quad [\text{V-s}]. \tag{2.5}$$

The left-hand integrals of the first two expressions are taken on the contour
surrounding the surface over which the right-hand integral is evaluated
(Figure 2.2). The last two expressions consider the surface surrounding a
closed volume.

Maxwell's equations are more general in their global form and remain
valid across boundaries. They can be used to solve problems that exhibit
circular, cylindrical, or spherical symmetry.

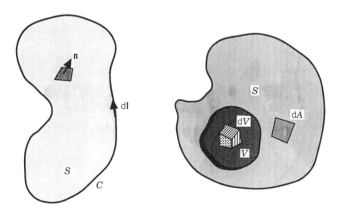

FIGURE 2.2   Volumes, surfaces, and contours of integration.

2.2 BOUNDARY CONDITIONS

When a structure consists of several homogeneous regions, for instance air and dielectric, the fields on both sides of the interface satisfy conditions on the interface between the two regions (Figure 2.3). These boundary conditions are derived from the global expression of Maxwell's equations (2.5) by integrating them over suitable contours, surfaces, and volumes, and taking the limit as $\delta \to 0$ [Ramo et al. 1965].

$$\mathbf{n} \times (\mathbf{E}_1 - \mathbf{E}_2) = 0, \qquad \mathbf{n} \cdot (\mathbf{D}_1 - \mathbf{D}_2) = \rho_s,$$
$$\mathbf{n} \times (\mathbf{H}_1 - \mathbf{H}_2) = \mathbf{J}_s, \qquad \mathbf{n} \cdot (\mathbf{B}_1 - \mathbf{B}_2) = 0. \tag{2.6}$$

where the vector $\mathbf{n}$ is perpendicular to the interface, pointing from region 2 toward region 1, $\mathbf{J}_s$ is the surface current density, and ρ_s is the surface charge density. The surface current and the surface charge densities, which are defined only on interfaces, should not be confused with the volume current and charge densities that appear in Maxwell's equations (2.4).

The tangential components of the electric field are continuous across any boundary, whereas those of the magnetic field become discontinuous when a surface current flows on it. When the conditions on the tangential components are satisfied, and the fields are solutions of Maxwell's equations on both sides, the normal field components also meet the boundary conditions.

2.3 MATERIAL PROPERTIES

2.3.1 Constitutive Equations: General Formulation

The two electric and the two magnetic quantities that appear in Maxwell's equations are related by constitutive equations that express the properties of the materials. In the most general case, they take the form

$$\mathbf{D} = \varepsilon_0 \mathbf{E} + \mathbf{P}, \qquad \mathbf{B} = \mu_0(\mathbf{H} + \mathbf{M}), \qquad \mathbf{J} = \sigma \mathbf{E}, \tag{2.7}$$

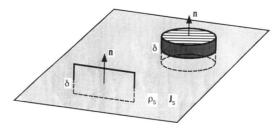

FIGURE 2.3 Interface between two media.

where ε_0 and μ_0 are, respectively, the electric and magnetic constants of a vacuum, and the material contributes the polarization field **P** and the magnetization field **M**. In most usual materials these two quantities are directly proportional to the fields when the amplitudes of the fields are small. Most engineering applications therefore assume that the materials are linear. At high field amplitudes, however, and in some materials the relationships are nonlinear, in which case the developments become quite involved. In most metals, the current density is directly proportional to the electric field through the conductivity σ.

2.3.2 Linear Isotropic Lossless Media

In a linear, isotropic, and lossless medium, **P** is directly proportional to the electric field **E**, while **M** is directly proportional to the magnetic field **H**. The constitutive expressions of Eq. 2.7 are simplified, yielding

$$\mathbf{D} = \varepsilon_0 \varepsilon_r \mathbf{E} = \varepsilon \mathbf{E}, \qquad \mathbf{B} = \mu_0 \mu_r \mathbf{H} = \mu \mathbf{H}. \tag{2.8}$$

They define, respectively, the relative and the absolute permittivities (ε_r and ε) and permeabilities (μ_r and μ) of the material. Most of the materials used to realize microstrip structures are nonmagnetic, so $\mu = \mu_0$ or $\mu_r = 1$, the main exception being the ferrite materials used to realize nonreciprocal devices, which are also anisotropic (Sections 2.3.3, 6.3.2, and 12.3).

2.3.3 Linear Anisotropic Lossless Media

In a linear, anisotropic, and lossless medium, the material properties depend on the direction and can be represented by tensors

$$\mathbf{D} = \varepsilon_0 \bar{\bar{\varepsilon}}_r \mathbf{E} = \bar{\bar{\varepsilon}} \mathbf{E}, \qquad \mathbf{B} = \mu_0 \bar{\bar{\mu}}_r \mathbf{H} = \bar{\bar{\mu}} \mathbf{H}. \tag{2.9}$$

The permittivity tensor and/or the permeability tensor are represented by 3×3 matrices. They are also called dyadics. The corresponding fields, **D** and **E** for an anisotropic permittivity, are then generally not collinear.

Dielectric anisotropy is encountered in woven and laminated composite substrate materials (Section 13.1.8) and in sapphire (Section 13.1.5). These materials possess one direction, the main axis, for which the permittivity is different from that in the perpendicular plane: they are called uniaxial.

Magnetic anisotropy is induced in ferrimagnetic materials by applying a biasing magnetic field. The material is then called gyrotropic, possessing unique nonreciprocal properties used to realize circulators and isolators (Section 6.3.2).

2.3.4 Linear Isotropic Lossy Media

In lossy materials, the physical processes require some time to get excited. The response follows the excitation with some delay (Figure 2.4). The retarding effect is mathematically expressed with convolution integrals:

$$\mathbf{D}(t) = \int_{-\infty}^{+\infty} \varepsilon(t - \tau)\mathbf{E}(\tau)\,\mathrm{d}\tau = \varepsilon(t) * \mathbf{E}(t),$$

$$\mathbf{B}(t) = \int_{-\infty}^{+\infty} \mu(t - \tau)\mathbf{H}(\tau)\,\mathrm{d}\tau = \mu(t) * \mathbf{H}(t). \qquad (2.10)$$

In the presence of losses, the displacement and the induction fields are functions of the electric and magnetic fields at all previous times (past history of the material). They are related through dielectric and magnetic impulse responses (Section 2.5.4). In the presence of material losses, the analysis of electromagnetic field interactions becomes very complicated and is seldom carried out directly in the time domain.

2.3.5 Perfect Electric Conductor

The conductivity σ in most metals is very large, meaning that the electric field within the conductor becomes vanishingly small since the current density $\mathbf{J}$ is finite. A simple approximation is obtained by letting $\sigma = \infty$, in

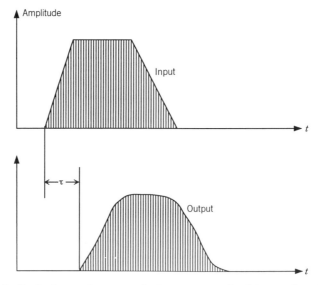

FIGURE 2.4 Excitation and response in the presence of a delay produced by losses.

which case one must have $\mathbf{E} = 0$ within the conductor: this is a *perfect electric conductor* (PEC).

On the boundary of a PEC, within which the field $\mathbf{E}_2 = 0$, Eq. 2.6 requires that

$$\mathbf{n} \times \mathbf{E} = 0. \tag{2.11}$$

This implies that the tangential component of the electric field vanishes on a PEC surface, so that the electric field $\mathbf{E}$ is everywhere normal to the boundary. The surface of a PEC generally supports an electric surface current $\mathbf{J}_s$, which accounts for the discontinuity in the tangential magnetic field (2.6).

2.3.6 Perfect Magnetic Conductor

The dual of the PEC is the *perfect magnetic conductor* (PMC), characterized by a vanishingly small internal magnetic field $\mathbf{H} \cong 0$, when the permeability μ of the material is extremely large. In ferromagnetic materials, the magnetic induction $\mathbf{B}$ must remain finite. On the boundary of a PMC, within which the magnetic field vanishes ($\mathbf{H}_2 = 0$) and on which no electric surface current flows, Eq. 2.6 specifies that

$$\mathbf{n} \times \mathbf{H} = 0. \tag{2.12}$$

This means that the tangential component of the magnetic field vanishes on a PMC: the magnetic field $\mathbf{H}$ is everywhere normal to the boundary. In practice, one seldom uses ferromagnetic materials to realize transmission lines, but the concept of the PMC remains useful to model open-circuit conditions and radiation through apertures.

2.3.7 Real Conductor

Metallic conductors possess a finite conductivity σ ($\neq \infty$) and thus dissipate some of the signal by Joule heating. At nearly all frequencies of interest in electrical engineering, the conduction current in a metal is very much larger than the displacement current, so

$$\sigma \gg \omega\varepsilon. \tag{2.13}$$

2.3.8 Superconductor

At very low temperatures some metals exhibit enhanced conduction properties (superconductivity), and anomalous skin effects were observed [Matick 1969]. Recently developed ceramic materials also exhibit superconducting behavior at much higher temperatures, above the boiling point of liquid nitrogen, and are called high-T_c superconductors. Since their discovery, great interest arose for the development of superconducting planar components. Very low loss implies low attenuation and low noise, but also smaller dispersion and the possibility of reducing the size of devices, leading to larger integration density and shorter interconnections.

The conduction mechanism in high-T_c superconductors is described by the two-fluid model with two kinds of charge carriers: superconducting electrons, grouped in Cooper pairs, and normal electrons. Conduction strongly depends on temperature and on current density: when a certain limiting value is reached, the superconducting effect decreases and then vanishes. This means that various parts of the same superconducting circuit may exhibit different conductivities [El-Ghazaly et al. 1992].

2.4   SINUSOIDAL TIME DEPENDENCE

2.4.1   Sine Waves

In their local form, Maxwell's equations include both space and time derivatives. They are thus partial derivative equations in four-dimensional space, with the fourth dimension (time) possessing characteristics different from the three space coordinates. Their resolution becomes much simpler when the time derivative can be removed.

In addition, the delay caused by losses (convolution integrals, Section 2.3.2) makes the time-domain analysis very difficult. The use of sine waves permits us to get rid of these cumbersome expressions.

One of the components of the electric field, E_x, having a sinusoidal time dependence of angular frequency $\omega = 2\pi f$ is expressed by

$$E_x(\mathbf{r}, t) = \sqrt{2}\, E_{x0}(\mathbf{r}) \cos(\omega t + \phi_x), \qquad (2.14)$$

where E_{x0} is the effective amplitude of the x component of the electric field, ϕ_x is its phase, and the vector $\mathbf{r}$ indicates the position in space. Similar expressions are obtained for the y and z components.

2.4.2   Complex Notation

The sine-wave dependence of time can be expressed in a compact manner by introducing complex notation. The electric field of (2.14) can be written as

$$E_x(\mathbf{r}, t) = \mathrm{Re}\left[\sqrt{2}\,\underline{E}_{x0}(\mathbf{r}) \exp(\mathrm{j}\omega t)\right] \qquad \text{with } \underline{E}_{x0}(\mathbf{r}) = E_{x0}(\mathbf{r}) \exp(\mathrm{j}\phi_x).$$
$$(2.15)$$

Proceeding in the same manner with the other two components and combining the three terms, one obtains

$$\mathbf{E}(\mathbf{r}, t) = \text{Re}\left[\sqrt{2}\,\underline{\mathbf{E}}(\mathbf{r})\exp(j\omega t)\right] \tag{2.16}$$

with

$$\underline{\mathbf{E}}(\mathbf{r}) = \mathbf{e}_x E_{x0}(\mathbf{r})\exp(j\phi_x) + \mathbf{e}_y E_{y0}(\mathbf{r})\exp(j\phi_y) + \mathbf{e}_z E_{z0}(\mathbf{r})\exp(j\phi_z).$$

The underlining indicates a complex quantity. Each field becomes a phasor–vector, that is a complex quantity made of two vectors: its real part is related to the field at time $t = 0$, the imaginary part at $t = -T/4$, where $T = 1/f$ is the period of the signal.

2.4.3 The Time Derivative

Taking the derivative of (2.16) with respect to time yields the multiplying factor $j\omega$. All time derivatives $\partial/\partial t$ in time-domain equations are replaced by the factor $j\omega$ in the frequency-domain notation.

2.4.4 The Constitutive Equations in Complex Notation

Delays and frequency-dependent material properties (dispersion) can readily be handled in the frequency domain, where the permittivity and the permeability become complex, frequency-dependent quantities. The constitutive relations in a linear isotropic lossy medium (Eq. 2.9) then become

$$\underline{\mathbf{D}} = \varepsilon_0 \underline{\varepsilon}_r \underline{\mathbf{E}} = \underline{\varepsilon}\underline{\mathbf{E}} = (\varepsilon' - j\varepsilon'')\underline{\mathbf{E}},$$

$$\underline{\mathbf{B}} = \mu_0 \underline{\mu}_r \underline{\mathbf{H}} = \underline{\mu}\underline{\mathbf{H}} = (\mu' - j\mu'')\underline{\mathbf{H}}. \tag{2.17}$$

The real parts of the permittivity and of the permeability express the reactive behavior of the material, while the imaginary parts are related to losses. The two are actually linked by the Kramers and Kronig relations, which indicate that the material is causal and has a bounded response, or, in other terms, that the response cannot antecede the excitation, and that the response to an excitation limited in time will eventually die out [Van Vleck 1951]. Here are the expressions for the permittivity:

$$\varepsilon'(\omega) = \text{Re}\left[\underline{\varepsilon}(\omega)\right] = \varepsilon_\infty + \frac{2}{\pi}\int_0^\infty \frac{\Omega\varepsilon''(\Omega)}{\Omega^2 - \omega^2}\,d\Omega,$$

$$\varepsilon''(\omega) = -\text{Im}\left[\underline{\varepsilon}(\omega)\right] = -\frac{2\omega}{\pi}\int_0^\infty \frac{\varepsilon''(\Omega) - \varepsilon_\infty}{\Omega^2 - \omega^2}\,d\Omega, \tag{2.18}$$

where the Cauchy principal part of the integral is defined by

$$\oint_0^\infty \frac{f(\Omega)}{\Omega^2 - \omega^2} \, d\Omega = \lim_{\delta \to 0} \int_0^{\omega - \delta} \frac{f(\Omega)}{\Omega^2 - \omega^2} \, d\Omega + \lim_{\delta \to 0} \int_{\omega + \delta}^{+\infty} \frac{f(\omega)}{\Omega^2 - \omega^2} \, d\Omega. \quad (2.19)$$

Equation 2.18 states that when a dielectric is dispersive (frequency dependent), it is also lossy, and vice versa. All dielectric materials are lossy somewhere within the frequency spectrum but not necessarily within the microwave range.

Similar relationships are established for the complex permeability.

2.4.5   Maxwell's Equations in Complex Notation

Maxwell's equations in phasor–vector notation take the form

$$\nabla \times \underline{\mathbf{E}} = -j\omega\mu\underline{\mathbf{H}}, \qquad \nabla \cdot \underline{\mathbf{E}} = \rho/\varepsilon,$$

$$\nabla \times \underline{\mathbf{H}} = j\omega\varepsilon\underline{\mathbf{E}} + \underline{\mathbf{J}}, \qquad \nabla \cdot \underline{\mathbf{H}} = 0. \quad (2.20)$$

With the time dependence removed, the equations become partial derivative equations in three-dimensional space.

2.4.6   Poynting Phasor – Vector: Power Density

Taking the cross product of the electric phasor vector by the complex conjugate of the magnetic phasor vector yields the Poynting phasor–vector $\underline{\mathbf{S}}$:

$$\underline{\mathbf{S}} = \underline{\mathbf{E}} \times \underline{\mathbf{H}}^* = \mathbf{P} + j\mathbf{Q} \quad [\text{W/m}^2]. \quad (2.21)$$

This quantity shows how electromagnetic power is transmitted by the waves. Its real part $\mathbf{P}$ is the active power density (that can be detected and used); its imaginary part $\mathbf{Q}$ corresponds to the reactive power (energy accumulation within the fields).

2.5   TIME-DOMAIN RESPONSES

A simple sinusoidal signal of frequency f carries power but no information. Any information-carrying signal must contain additional time variations of amplitude, frequency, or phase, called modulations. An information-carrying signal is thus always more complicated than a simple sine wave.

2.5.1   Fourier Transform, Frequency Spectrum

Any time-varying signal can be developed over a sum of sine and cosine functions of time or of complex exponential functions, with different frequencies, that form its *spectrum* in the frequency domain. The spectrum of a periodical signal is made of discrete lines and is called the Fourier series. Nonperiodical signals possess a continuous spectrum in the frequency domain, provided by their *Fourier transform*, defined for a signal $f(t)$ by [Bracewell 1978]

$$\underline{F}(\omega) = \int_{-\infty}^{+\infty} f(t) \exp(-j\omega t)\, dt. \tag{2.22}$$

The inverse Fourier transform, which determines the time function $f(t)$ in terms of its spectrum $\underline{F}(\omega)$ is then similarly given by

$$f(t) = \frac{1}{2\pi} \int_{-\infty}^{+\infty} \underline{F}(\omega) \exp(+j\omega t)\, d\omega. \tag{2.23}$$

2.5.2   Some Basic Properties

The function $f(t)$ represents any physical parameter, so it is always a *real quantity*. This imposes a condition on $\underline{F}(\omega)$ that can be deduced from (2.20):

$$\underline{F}(-\omega) = \underline{F}^*(\omega). \tag{2.24}$$

The complex function $\underline{F}(\omega)$ is called *Hermitian*.

The pair of functions defined by Eqs. 2.22 and 2.23 provides a way to move back and forth between the time domain and the frequency domain. The Fourier transforms of some simple functions of time are presented in Table 2.1. The transform of an infinitely short unit Dirac pulse $\delta(t)$ is simply unity, whereas the transform of the Heaviside unit step $\mathbf{1}(t)$ is $1/j\omega$. Transforms of periodic functions like $\sin \omega_0 t$, $\cos \omega_0 t$ contain Dirac delta functions in the frequency domain (discrete line spectrum).

In the frequency domain, the derivation and the integration with respect to time are replaced by multiplication and division, respectively, by the factor $j\omega$ (Section 2.4.3). A delay τ in the time domain is replaced by a multiplication by the term $\exp(-j\omega\tau)$. The transforms of other functions of time may be obtained by combining functions listed in Table 2.1. For instance, a rectangular pulse of amplitude A and of duration τ_0 is the combination of two Heaviside unit steps (Figure 2.5):

$$f(t) = A[\mathbf{1}(t) - \mathbf{1}(t - \tau_0)], \tag{2.25}$$

TABLE 2.1 Fourier Transforms

$f(t)$	$\underline{F}(j\omega)$
$\delta(t)$	1
$\delta(t-a)$	$\exp(-j\omega a)$
$1(t)$	$\dfrac{1}{j\omega}$
$1(t-a)$	$\dfrac{\exp(-j\omega a)}{j\omega}$
$t1(t)$	$-\dfrac{1}{\omega^2}$
$\exp(at)\,1(t)$	$\dfrac{1}{j\omega - a}$
$\sin(bt)\,1(t)$	$\dfrac{b}{b^2 - \omega^2}$
$\cos(bt)\,1(t)$	$\dfrac{j\omega}{b^2 - \omega^2}$
$\exp(at)\sin(bt)\,1(t)$	$\dfrac{b}{b^2 + (j\omega - a)^2}$
$\exp(at)\cos(bt)\,1(t)$	$\dfrac{j\omega - a}{b^2 + (j\omega - a)^2}$
$J_0(2\sqrt{at}\,)\,1(t)$	$\dfrac{\exp(-a/j\omega)}{j\omega}$
$\mathrm{erfc}\left(\dfrac{a}{2}\sqrt{t}\,\right)1(t)$	$\dfrac{\exp(-a\sqrt{j\omega}\,)}{j\omega}$

where

$$1(t) = \begin{cases} 0 & \text{for } t < 0, \\ 1 & \text{for } t > 0. \end{cases}$$

Its transform is given by

$$\underline{F}(\omega) = (A/j\omega)\left[1 - \exp(-j\omega\tau_0)\right]. \tag{2.26}$$

The response to a rectangular pulse can thus be obtained by combining the responses to two unit step functions.

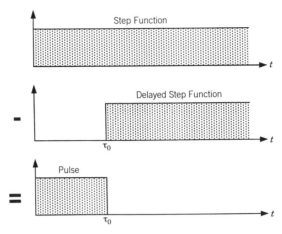

FIGURE 2.5   Pulse of finite duration.

2.5.3   Response to an Excitation

The response $g(t)$ to any arbitrary excitation $f(t)$ is obtained by the following steps:

1. The excitation $f(t)$ is expressed in the frequency domain by taking its Fourier transform $\underline{F}(\omega)$, defined by Eq. 2.22.

2. The transfer function $\underline{H}(\omega)$ at the angular frequency ω is the ratio of the transform of the response signal $\underline{G}(\omega)$ to that of the excitation $\underline{F}(\omega)$:

$$\underline{H}(\omega) = \underline{G}(\omega)/\underline{F}(\omega). \qquad (2.27)$$

It is given by the analysis of the microstrip circuit in the frequency domain (complex notation).

3. The response transform $\underline{G}(\omega)$ is then provided by

$$\underline{G}(\omega) = \underline{H}(\omega)\underline{F}(\omega) = \underline{H}(\omega) \int_{-\infty}^{+\infty} f(\tau) \exp(-j\omega\tau)\, d\tau. \qquad (2.28)$$

4. Taking the inverse Fourier transform of $\underline{G}(\omega)$ (Eq. 2.22) then yields the response $g(t)$ in the time domain:

$$
\begin{aligned}
g(t) &= \frac{1}{2\pi} \int_{-\infty}^{+\infty} \underline{G}(\omega) \exp(+j\omega t)\, d\omega \\
&= \frac{1}{2\pi} \int_{-\infty}^{+\infty} \underline{H}(\omega) \int_{-\infty}^{+\infty} f(\tau) \exp(-j\omega\tau) \exp(+j\omega t)\, d\tau\, d\omega. \quad (2.29)
\end{aligned}
$$

This expression is further developed mathematically to obtain

$$g(t) = \frac{1}{2\pi} \int_{-\infty}^{+\infty} \underline{H}(\omega) \int_{-\infty}^{+\infty} f(\tau) \exp[+j\omega(t-\tau)] \, d\tau \, d\omega. \quad (2.30)$$

Introducing the impulse response $h(t)$ of the system, expressed by

$$\underline{H}(\omega) = \int_{-\infty}^{+\infty} h(t) \exp(-j\omega t) \, dt, \quad (2.31)$$

we then obtain

$$g(t) = \frac{1}{2\pi} \int_{-\infty}^{+\infty} \int_{-\infty}^{+\infty} \int_{-\infty}^{+\infty} f(\tau) h(u) \exp[+j\omega(t-\tau-u)] \, d\tau \, du \, d\omega. \quad (2.32)$$

A change of variables yields

$$g(t) = \frac{1}{2\pi} \int_{-\infty}^{+\infty} \int_{-\infty}^{+\infty} \int_{-\infty}^{+\infty} f(\tau) h(u-\tau) \exp[+j\omega(t-u)] \, d\tau \, du \, d\omega. \quad (2.33)$$

Finally, carrying out the integration with respect to ω and to u, we have

$$g(t) = \int_{-\infty}^{+\infty} \int_{-\infty}^{+\infty} f(\tau) h(u-\tau) \delta(t-u) \, d\tau \, du$$

$$= \int_{-\infty}^{+\infty} f(\tau) h(t-\tau) \, d\tau. \quad (2.34)$$

This is the convolution of the two functions $f(t) * h(t)$.

2.5.4   The Dielectric Response

The relationship between the electric field in a lossy material and the resulting displacement field, as given in Eq. 2.9, is one particular response in the time domain. The dielectric impulse response is the inverse Fourier transform of the complex permittivity:

$$\varepsilon(t) = \frac{1}{2\pi} \int_{-\infty}^{+\infty} \underline{\varepsilon}(\omega) \exp(+j\omega t) \, d\omega. \quad (2.35)$$

2.6   POTENTIALS AND WAVE EQUATIONS

2.6.1   Wave Equations for the Fields

The combination of the two Maxwell curl equations of Eq. 2.20 yields second-order wave equations involving only one field:

$$\nabla^2 \underline{E} + \omega^2 \underline{\varepsilon}\underline{\mu}\underline{E} = \nabla\underline{\rho}/\underline{\varepsilon} + j\omega\underline{\mu}\underline{J},$$

$$\nabla^2 \underline{H} + \omega^2 \underline{\varepsilon}\underline{\mu}\underline{H} = -\nabla \times \underline{J}. \tag{2.36}$$

The electric current and electric charge densities are the sources of the fields. It is apparent that, while the magnetic field is produced by the sole electric current, the electric field results from both the current and the charge. When the materials considered are lossless, the solutions are waves that travel with velocity $1/\sqrt{\varepsilon\mu}$.

2.6.2   Vector and Scalar Potentials

It is here possible to define potentials, respectively a magnetic vector potential and an electric scalar potential:

$$\underline{B} = \nabla \times \underline{A} \qquad \text{and} \qquad \underline{E} + j\omega\underline{\mu}\underline{A} = -\nabla\underline{V}. \tag{2.37}$$

The Lorentz gauge is generally introduced at this point, specifying a particular relationship between the two potentials:

$$\nabla \cdot \underline{A} + j\omega\underline{\varepsilon}\underline{\mu}\underline{V} = 0. \tag{2.38}$$

2.6.3   Wave Equations for the Potentials

The wave equations for the complex potentials then yield

$$\nabla^2 \underline{A} + \omega^2 \underline{\varepsilon}\underline{\mu}\underline{A} = -\underline{\mu}\underline{J},$$

$$\nabla^2 \underline{V} + \omega^2 \underline{\varepsilon}\underline{\mu}\underline{V} = -\underline{\rho}/\underline{\varepsilon}. \tag{2.39}$$

The source of the magnetic vector potential is the electric current, while that of the electric scalar potential is the charge density. There are here four differential equations, instead of six in Eq. 2.36.

2.6.4   Plane Waves in Free Space

Solving this expression in a rectangular coordinate system yields plane waves, whose vector potential has a spatial dependence of the form

$$\underline{\mathbf{A}} = \underline{\mathbf{A}}_0 \exp(-\underline{\gamma} \cdot \mathbf{r}) \tag{2.40}$$

with

$$\underline{\gamma} \cdot \underline{\gamma} = -\omega^2 \underline{\varepsilon}\mu. \tag{2.41}$$

Plane waves can either be uniform, in which case the two vector components (real and imaginary parts) of $\underline{\gamma}$ point along the same direction (radiated waves, Section 1.3.3), or still one of them may vanish. They can also be nonuniform, when the two vector components are not collinear (surface waves, Section 1.3.5, and leaky waves, Section 1.3.4).

Plane waves provide an interesting approximation to study wave propagation, even though, strictly speaking, they do not exist in the real world.

2.6.5   Plane Waves in a Metal

Solving Maxwell's equation for uniform plane waves within a metal yields a complex propagation factor $\underline{\gamma}$ and a complex metal impedance $\underline{Z}_m$:

$$\underline{\gamma} = \alpha + \mathrm{j}\beta = (1 + \mathrm{j})\sqrt{\frac{\omega\mu\sigma}{2}} = \frac{1 + \mathrm{j}}{\delta}, \tag{2.42}$$

$$\underline{Z}_m = \sqrt{\frac{\mathrm{j}\omega\mu}{\sigma}} = (1 + \mathrm{j})\sqrt{\frac{\omega\mu}{2\sigma}} = (1 + \mathrm{j})R_m. \tag{2.43}$$

The skin depth $\delta = 1/\alpha$ represents the distance traveled by a wave into a material until its amplitude decreases by a factor $\mathrm{e} \cong 2.718$.

2.6.6   Plane Waves in a Superconductor

In a superconductor, the conductivity becomes a complex quantity $\sigma_{sc} = \sigma_1 + \mathrm{j}\sigma_2$, where both σ_1 and σ_2 are functions of the temperature [Baiocchi 1992].

2.6.7   Spherical Waves in Free Space

Waves are actually excited by localized sources, so radiated fields are solutions of the wave equation in spherical coordinates. The magnetic vector potential then takes the form

$$\underline{\mathbf{A}} = \underline{\mathbf{A}}_0 \frac{\exp\left(-\mathrm{j}\omega\sqrt{\underline{\varepsilon}\mu}\, r\right)}{r}. \tag{2.44}$$

This is the response to an infinitesimal element of current (Hertz dipole) located at the origin of the coordinate system ($r = 0$), at which point the potential becomes singular with a simple pole. Higher-order poles denote the presence of higher-order current sources (dipoles, quadrupoles, and so on). Any field distribution can be built up by combining multipoles, located either at the origin or at other carefully selected locations [Hafner 1990].

2.6.8 Green's Functions

In all linear structures, the general solution can be expressed by a superposition of partial solutions, and the vector potential created by any given set of currents is then provided by an integral involving all the current sources:

$$\underline{A}(\mathbf{r}) = \frac{\mu_0}{4\pi} \int_{V'} dV \frac{\mathbf{J}(\mathbf{r'}) \exp\left(-j\omega\sqrt{\varepsilon\mu}\,|\mathbf{r} - \mathbf{r'}|\right)}{|\mathbf{r} - \mathbf{r'}|}$$

$$= \int_{V'} dV\, \underline{\mathbf{J}}(\mathbf{r'}) G_A(\mathbf{r} - \mathbf{r'}). \tag{2.45}$$

This is a convolution integral in the space domain $\mathbf{J}(\mathbf{r}) * G_A$. The coordinate $\mathbf{r'}$ corresponds to the currents (sources) located within volume V', while the unprimed coordinate $\mathbf{r}$ is linked to the observer. When the vector potential has been determined everywhere in space, the electric and magnetic fields can be derived from it. The term G_A is the free-space Green's function.

2.6.9 Comparison

The two previous sections present direct similarities between the time response and the spatial dependence. Comparing Eqs. 2.34 and 2.45 shows that the Green's function is the "impulse response" in the space domain.

2.7 PROBLEMS

2.7.1 Determine the force produced by an electric field of 100 V/m acting on a charge of 5 coulombs [A-s].

2.7.2 The electric field of a wave is given by the expression

$$\mathbf{E} = \mathbf{e}_x \sin(\omega t - \beta y) \qquad \text{where } \omega \text{ and } \beta \text{ are constants.}$$

Determine the corresponding magnetic field in air, assuming that it has a value of zero for $t = 0$ and $y = 0$. Determine also the electric charge density.

2.7.3   The electric field of a wave is given by the expression

$$\mathbf{E} = \mathbf{e}_x \sin(\omega t - \beta x) \qquad \text{where } \omega \text{ and } \beta \text{ are constants.}$$

Determine the corresponding magnetic field in air, assuming that it has a value of zero for $t = 0$ and $y = 0$. Determine also the electric charge density.

2.7.4   Determine whether the following expression can represent an induction field:

$$\mathbf{B} = \mathbf{e}_z \sin \omega t \cos \frac{\pi x}{a} \cos \frac{\pi z}{b},$$

where a and b are constants.

2.7.5   At which frequency is the conduction current in copper equal to the displacement current (Eq. 2.13)? The relative permittivity is assumed to equal unity.

2.7.6   Find out and sketch the behavior over one period of the electric field corresponding to the phasor vector $\underline{\mathbf{E}} = \mathbf{e}_z + j\mathbf{e}_x$.

2.7.7   Determine from which vector potentials the induction field $\mathbf{B} = yz\,\mathbf{e}_z - xy\,\mathbf{e}_x$ can derive.

2.7.8   Determine under which conditions plane waves traveling in air can be nonuniform.

2.7.9   Determine the skin depth and the metal impedance in copper at 30 GHz.

CHAPTER THREE

Transmission Lines

A transmission line is made of two or more conducting strips or wires, along which electrical signals propagate, ideally without attenuation. This implies that it is made of low-loss materials—good conductors (metals) and good insulators (dielectrics). Theoretical developments, considering most often lossless lines, provide approximate values for the parameters. The attenuation and the distortion produced by losses in actual materials can be determined by perturbation [Morse and Feshbach 1953]. The onset of radiation limits the frequency range of microstrip lines.

3.1 HOMOGENEOUS UNIFORM LINE

3.1.1 Definition of the Structure

A simple transmission line, made of two straight metallic conductors embedded in a homogeneous lossless dielectric (Figure 3.1), is considered in this section. The conductors are infinitely long, parallel to the longitudinal axis z. The cross section of the line and its material parameters are independent of the longitudinal position: shifting along the z direction does not modify the structure, which is called *uniform* or *translation invariant*.

The coordinate system contains the longitudinal coordinate z and two coordinates in the transverse plane (perpendicular to z), which are the rectangular coordinates x (tangential to the ground plane) and y (normal to the ground plane). The transverse dimensions (substrate height, strip width) are much smaller than a wavelength.

3.1.2 Separation of Variables

Since the transverse plane is perpendicular to the direction of propagation z (Figure 3.1), transverse and longitudinal field dependencies are independent

33

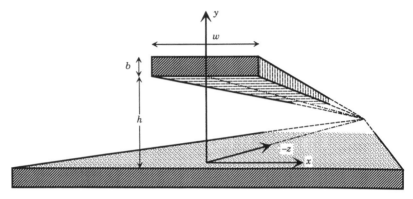

FIGURE 3.1   Uniform homogeneous line.

of each other, or *orthogonal*, and may be analyzed separately. The differential operator ∇ that appears in Maxwell's equations (2.4) is separated into two parts:

$$\nabla = \nabla_t + \mathbf{e}_z \frac{\partial}{\partial z}. \tag{3.1}$$

The first term on the right-hand side involves transverse dependence, while the second one relates only to longitudinal dependence. Similarly, the vector **r**, which defines position, has transverse and longitudinal parts:

$$\mathbf{r} = \mathbf{r}_t + \mathbf{e}_z z. \tag{3.2}$$

The same is true for every field (Eq. 2.1):

$$\mathbf{X} = \mathbf{X}_t + \mathbf{e}_z X_z. \tag{3.3}$$

The wave equations (Section 2.6) are then separated into transverse and longitudinal parts and analyzed independently [Gardiol 1987].

The boundaries are assumed to be made of perfect electric conductors (PEC, Section 2.3.5). The tangential components of the electric field vanish on their surface (Eq. 2.11), where the electric current is concentrated.

3.1.3   Modes of Propagation

Several field configurations, called modes of propagation, satisfy Maxwell's equations in the presence of the transverse boundary conditions of the line. Every mode possesses its own propagation characteristics: attenuation and phase shift per unit length, propagation velocities, and cutoff frequency. The modes form a discrete set when the structure is surrounded by a metal boundary, to which must be added a continuous spectrum of radiating modes when the structure is open.

Different propagating modes travel at different velocities, so the signal excited by a source gets distorted. Selecting a low enough signal frequency permits only one mode to propagate on the line (the dominant mode), and thus multimode distortion is prevented. The excitation of propagating higher-order modes sets an upper limit to the operating frequency range of transmission lines.

The dominant mode of a two-conductor line (Figure 3.1) has no cutoff, so a signal at any frequency can propagate along the line. When the structure is also homogeneous and the conductor lossless, the signal is not attenuated or distorted. Both its electric and magnetic fields are transverse ($H_z = 0$ and $E_z = 0$), and the mode is called *transverse electromagnetic* (TEM).

3.1.4 TEM Field Analysis

The fields of a transverse electromagnetic wave are both transverse to the direction of propagation, implying that the longitudinal components of the fields vanish everywhere:

$$E_z = 0 \qquad \text{and} \qquad H_z = 0. \qquad (3.4)$$

The only nonzero components are the transverse ones (perpendicular to the direction of propagation), and thus

$$\mathbf{E} = \mathbf{E}_t \qquad \text{and} \qquad \mathbf{H} = \mathbf{H}_t. \qquad (3.5)$$

The subscripts t for the field components are thus superfluous and will be left out. Separating the longitudinal and the transverse parts of the operator ∇ (Eq. 3.1) in Maxwell's divergence equations (2.17) yields

$$\nabla_t \cdot \underline{\mathbf{E}} = 0 \qquad \text{and} \qquad \nabla_t \cdot \underline{\mathbf{H}} = 0. \qquad (3.6)$$

Note that there are no free charges within a dielectric (i.e., $\rho = 0$). The curl equations (2.17) are similarly developed:

$$\left(\nabla_t + \mathbf{e}_z \frac{\partial}{\partial z} \right) \times \underline{\mathbf{E}} = \nabla_t \times \underline{\mathbf{E}} + \mathbf{e}_z \frac{\partial}{\partial z} \times \underline{\mathbf{E}} = -j\omega\mu \underline{\mathbf{H}},$$

$$\left(\nabla_t + \mathbf{e}_z \frac{\partial}{\partial z} \right) \times \underline{\mathbf{H}} = \nabla_t \times \underline{\mathbf{H}} + \mathbf{e}_z \frac{\partial}{\partial z} \times \underline{\mathbf{H}} = -j\omega\varepsilon \underline{\mathbf{E}}. \qquad (3.7)$$

The two expressions are separated into their transverse and longitudinal parts, yielding

transverse parts

$$\frac{\partial}{\partial z}\underline{\mathbf{E}} = j\omega\mu\mathbf{e}_z \times \underline{\mathbf{H}} \quad \text{and} \quad \frac{\partial}{\partial z}\underline{\mathbf{H}} = -j\omega\varepsilon\mathbf{e}_z \times \underline{\mathbf{E}}; \quad (3.8)$$

longitudinal parts

$$\nabla_t \times \underline{\mathbf{E}} = 0 \quad \text{and} \quad \nabla_t \times \underline{\mathbf{H}} = 0. \quad (3.9)$$

3.1.5 Wave Propagation

Combining the two transverse parts (3.8) yields the wave equations

$$\frac{\partial^2}{\partial z^2}\underline{\mathbf{E}} + \omega^2\varepsilon\mu\underline{\mathbf{E}} = 0 \quad \text{or} \quad \frac{\partial^2}{\partial z^2}\underline{\mathbf{H}} + \omega^2\varepsilon\mu\underline{\mathbf{H}} = 0. \quad (3.10)$$

The solutions are exponentials having the general form

$$\underline{\mathbf{E}} = \underline{\mathbf{E}}_+ \exp\left(-j\omega\sqrt{\varepsilon\mu}\, z\right) + \underline{\mathbf{E}}_- \exp\left(+j\omega\sqrt{\varepsilon\mu}\, z\right), \quad (3.11)$$

$$\underbrace{\phantom{\underline{\mathbf{E}}_+ \exp\left(-j\omega\sqrt{\varepsilon\mu}\, z\right)}}_{\text{forward wave}} \qquad \underbrace{\phantom{\underline{\mathbf{E}}_- \exp\left(+j\omega\sqrt{\varepsilon\mu}\, z\right)}}_{\text{reverse wave}}$$

where $\underline{\mathbf{E}}_+$ and $\underline{\mathbf{E}}_-$ are transverse phasor–vectors, whose amplitude and phase are specified by the boundaries at the ends of the line (generator and load). The propagation characteristics are those of a uniform plane wave within the same medium (Section 2.6.4) and are not affected by the metal boundaries.

3.1.6 Laplace's Equations

Since the transverse curls of the electric and of the magnetic field both vanish, the fields derive from transverse potentials:

$$\underline{\mathbf{E}} = -\nabla_t\underline{\Phi} \quad \text{and} \quad \underline{\mathbf{H}} = -\nabla_t\underline{\Theta}. \quad (3.12)$$

Applying the transverse divergence equations (3.6) yields

$$\nabla_t^2\underline{\Phi} = 0 \quad \text{and} \quad \nabla_t^2\underline{\Theta} = 0. \quad (3.13)$$

These are two-dimensional Laplace's equations, to be solved in the presence of the boundary conditions. In any transverse plane, the electric potential $\underline{\Phi}$ is constant on the conductors, $\underline{\Phi}$ is generally set equal to zero on the large conductor (ground plane), and $\underline{\Phi} = \underline{U}$ on the smaller strip (upper conductor).

Many techniques are available to solve Laplace's equation in two-dimensional electrostatics [Plonsey and Collin 1961]. The most usual ones are conformal mapping in the case of open microstrips (Section 10.2) and approximate techniques such as finite differences (Section 10.4) [Forsythe and Wasow 1960] and finite elements (Section 10.5.4) [Strang and Fix 1973] when the microstrip is enclosed in a metal box.

3.2   CURRENTS AND VOLTAGES ON A LINE

3.2.1   Line Current and Voltage

In circuit analysis, voltages and currents are preferred to electromagnetic fields. In a two-conductor line, the transverse voltage is the line integral of the electric field between the two conductors:

$$\underline{U}(z) = -\int_A^B \underline{\mathbf{E}}(\mathbf{r}, z) \cdot \mathbf{dl}. \tag{3.14}$$

The path of integration is shown in Figure 3.2.

The longitudinal current on the strip is the integral of the surface current density around the strip, making use of the boundary conditions on a perfect electric conductor:

$$\underline{I}(z) = \oint_C \underline{\mathbf{H}}(\mathbf{r}, z) \cdot \mathbf{dl}. \tag{3.15}$$

3.2.2   Power Transfer

The power carried by the transmission line is the integral of the longitudinal component of the Poynting vector (Section 2.4.6) over the cross section of the

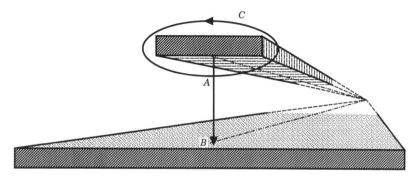

FIGURE 3.2   Integration paths and contours on microstrip.

line. For microstrip, the cross section extends over the complete upper half of the transverse plane. The real part is the active power; the imaginary part is the reactive power.

Carrying out the integration for a homogeneous line, the power becomes the product of the voltage by the complex conjugate of the current [Gardiol 1987]:

$$\underline{S}(z) = \underline{U}(z)\underline{I}^*(z) = P(z) + jQ(z). \tag{3.16}$$

3.2.3 Capacitance and Inductance

The capacitance $C'\,dz$ is the ratio of the charge stored on a strip of infinitesimal length dz to the transverse voltage. The charge density is obtained from the boundary conditions (2.6) and then integrated over the contour of the strip, yielding

$$C'\,dz = -\frac{\varepsilon\,dz\,\oint_C \mathbf{e}_z \times \underline{\mathbf{E}}(\mathbf{r},z)\cdot d\mathbf{l}}{\int_A^B \underline{\mathbf{E}}(\mathbf{r},z)\cdot d\mathbf{l}}. \tag{3.17}$$

In a similar way, the inductance $L'\,dz$ is the ratio of the magnetic flux crossing a section of infinitesimal length dz between the two conductors to the line current:

$$L'\,dz = \frac{\mu\,dz\,\int_A^B \mathbf{e}_z \times \underline{\mathbf{H}}(\mathbf{r},z)\cdot d\mathbf{l}}{\oint_C \underline{\mathbf{H}}(\mathbf{r},z)\cdot d\mathbf{l}}. \tag{3.18}$$

Here, C' and L' are, respectively, the capacitance and the inductance per unit length of the transmission line.

3.2.4 Transmission Line Equations

Integrating the first equation in (3.8) from A to B in Figure 3.2 yields

$$\frac{\partial}{\partial z}\int_A^B \underline{\mathbf{E}}(\mathbf{r},z)\cdot d\mathbf{l} = j\omega\mu\int_A^B \mathbf{e}_z \times \underline{\mathbf{H}}(\mathbf{r},z)\cdot d\mathbf{l}. \tag{3.19}$$

The introduction of the values for current, voltage, and inductance per unit length yields the first line equation

$$\frac{d\underline{U}(z)}{dz} = -j\omega L'\underline{I}(z). \tag{3.20}$$

Similarly, integrating the second equation of (3.8) over the contour of the

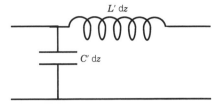

FIGURE 3.3 Equivalent circuit for a length dz of transmission line.

strip yields

$$\frac{\partial}{\partial z}\oint_C \underline{H}(\mathbf{r}, z) \cdot d\mathbf{l} = -j\omega\varepsilon\oint_C \mathbf{e}_z \times \underline{E}(\mathbf{r}, z) \cdot d\mathbf{l}, \qquad (3.21)$$

from which one obtains the second line equation

$$\frac{d\underline{I}(z)}{dz} = -j\omega C'\underline{U}(z). \qquad (3.22)$$

An infinitesimal section of transmission line can be represented by the equivalent circuit of Figure 3.3.

3.2.5   Wave Propagation along a Transmission Line

Combining the two transmission line equations (3.20) and (3.22) yields a second-order wave equation involving either only voltage or only current:

$$\frac{d^2\underline{U}(z)}{dz^2} = -\omega^2 L'C'\underline{U}(z) \quad \text{or} \quad \frac{d^2\underline{I}(z)}{dz^2} = -\omega^2 L'C'\underline{I}(z). \qquad (3.23)$$

Comparing these equations with the wave equation for the fields (Eq. 3.10) and noting that voltage and current are related to the fields, it is apparent that

$$L'C' = \varepsilon\mu = 1/c^2, \qquad (3.24)$$

where c is the velocity of the wave in the homogeneous medium ε, μ.
  The solutions to the equations are exponentials of the form

$$\underline{U}(z) = \underset{\text{forward wave}}{\underline{U}_+ \exp(-j\beta z)} + \underset{\text{reverse wave}}{\underline{U}_- \exp(+j\beta z)},$$

$$\underline{I}(z) = \underset{\text{forward wave}}{Y_c\underline{U}_+ \exp(-j\beta z)} - \underset{\text{reverse wave}}{Y_c\underline{U}_- \exp(+j\beta z)}, \qquad (3.25)$$

where the propagation factor is $\beta = \omega\sqrt{L'C'} = \omega\sqrt{\varepsilon\mu}$ (phase shift per unit length of line) and $Y_c = (C'/L')^{1/2}$ is the characteristic admittance. By Eq. 3.24, Y_c takes the forms

$$Y_c = \frac{1}{Z_c} = \sqrt{\frac{C'}{L'}} = cC' = \frac{1}{cL'}. \tag{3.26}$$

3.3 UNIFORM MICROSTRIP LINE

3.3.1 Definition of the Structure

The fields in a real microstrip line (Figure 3.4) extend within two media, air above and dielectric below, so that the structure is inhomogeneous. A uniform microstrip line has infinitely long conductors, forming a translation invariant structure.

3.3.2 Hybrid Modes of Propagation

For a TEM wave, the propagation velocity depends only on the material properties ε and μ. Within two materials, TEM waves propagate with two velocities, one within the dielectric substrate, the other in the air above the interface. The boundary conditions on the interface require the continuity of the tangential components and cannot be satisfied. As a result, the propagation along a microstrip line cannot be pure TEM.

Considering Maxwell's equations and boundary conditions, it was shown that the longitudinal components of the electric and of the magnetic field of the dominant mode do not vanish (Figure 3.5) [Gardiol 1984]. The dominant mode is therefore a hybrid mode.

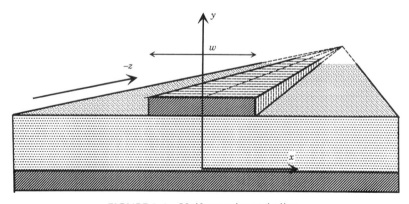

FIGURE 3.4   Uniform microstrip line.

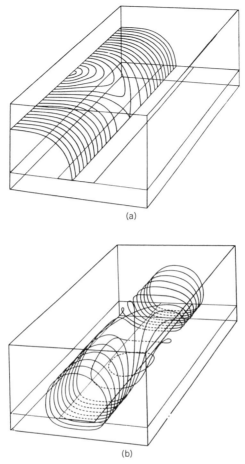

(a)

(b)

FIGURE 3.5 Electric and magnetic field lines on an enclosed microstrip. (Courtesy G. K. Grünberger 1971).

3.3.3 Power on a Microstrip Line

The complex power $\underline{S}$ carried along an inhomogeneous microstrip, given by the integration of the longitudinal component of Poynting's vector, is no longer given by Eq. 3.17, simply taking the product $\underline{U}\underline{I}^*$. This is due to the presence of the longitudinal components of the fields, and the same situation is encountered in waveguides [Schelkunoff 1943].

It is then possible to define three characteristic impedances by considering the three pairs of terms that can be formed with $\underline{U}$, $\underline{I}$, and $\underline{S}$:

$$\underline{Z}_{UI} = \frac{\underline{U}}{\underline{I}}, \qquad \underline{Z}_{PI} = \frac{\underline{S}}{|\underline{I}|^2}, \qquad \underline{Z}_{PU} = \frac{|\underline{U}|^2}{\underline{S}}. \qquad (3.27)$$

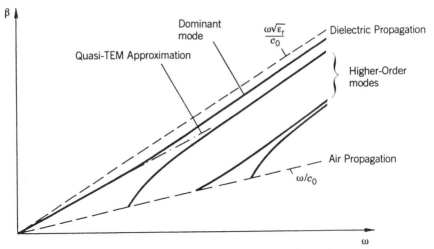

FIGURE 3.6 Dispersion diagram of a microstrip.

The three impedances take the same value at the low-frequency limit, and the differences between them increase with frequency as the longitudinal components become significant. It is only when the dispersion becomes important (Section 3.5) that the three values differ markedly from each other.

3.3.4 Dispersion Diagram

Due to the triply inhomogeneous nature of microstrip (Section 1.5.2), the analysis of the electromagnetic fields is difficult, and many different approaches were considered to carry out this task (Chapter 10). The propagation factor β of the dominant mode is not a linear function of frequency (Figure 3.6), except at the low-frequency end of the range.

3.4 QUASI-TEM APPROXIMATION

Over most of the operating frequency range of microstrips, the longitudinal components of the fields for the dominant mode remain very much smaller than the transverse components and may therefore be neglected. The dominant mode then behaves like a TEM mode, and the developments of Sections 3.1 and 3.2 remain valid for its analysis: this is called the quasi-TEM approximation.

3.4.1 Effective Permittivity

In the quasi-TEM approximation, the inhomogeneous microstrip is replaced by a homogeneous structure, in which the conductors retain the same

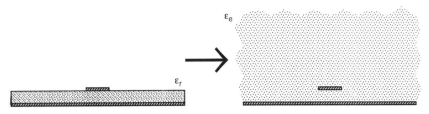

FIGURE 3.7 Principle of the quasi-TEM approximation.

geometry (w, h, b) but are immersed in a single dielectric of effective permittivity ε_e (Figure 3.7).

The low-frequency value of ε_e is determined by evaluating the capacitance of the inhomogeneous structure, yielding approximate relationships [Hammerstad and Jansen 1980]

$$\varepsilon_e \cong \frac{\varepsilon_r + 1}{2} + \frac{\varepsilon_r - 1}{2}\left(1 + 10\frac{h}{w}\right)^{-ab}, \tag{3.28}$$

with

$$a = 1 + \frac{1}{49}\log\frac{(w/h)^4 + (w/52h)^2}{(w/h)^4 + 0.432}$$

$$+ \frac{1}{18.7}\log\left\{1 + \left(\frac{1}{18.1}\frac{w}{h}\right)^3\right\},$$

$$b = 0.564\left(\frac{\varepsilon_r - 0.9}{\varepsilon_r + 3}\right)^{0.053}.$$

The accuracy provided by this approximation is better than 0.2% for $0.01 \le w/h \le 100$ and $1 \le \varepsilon_r \le 128$.

3.4.2 Characteristic Impedance

To determine the characteristic impedance Z_c or its inverse, the characteristic admittance Y_c, one must determine either C' or L' from Eqs. 3.16 or 3.17. When the strip is infinitely thin ($b = 0$), the fields can be determined exactly by conformal mapping [Schneider 1969]. This technique does not yield an explicit expression for the characteristic impedance, so an approximate formulation for Z_c was established by curve-fitting techniques. An accuracy of 0.03% over the range $0 \le w/h \le 1000$ was reported by Hammerstad and

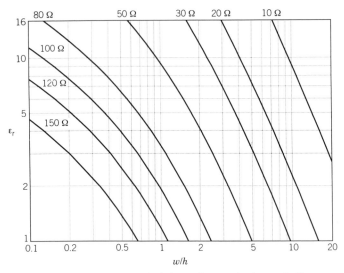

FIGURE 3.8 Characteristic impedance of microstrip line.

Jansen [1980] for the expression

$$Z_c \cong \frac{1}{2\pi} \sqrt{\frac{\mu_0}{\varepsilon_0 \varepsilon_e}} \log\left\{ F_1 \frac{h}{w} + \sqrt{1 + \left(2\frac{h}{w}\right)^2} \right\}, \qquad (3.29)$$

with $F_1 = 6 + (2\pi - 6)\exp\{-(30.666h/w)^{0.7528}\}$.

The characteristic impedance is a function of the geometric ratio w/h and the effective relative permittivity ε_e. The characteristic impedance is shown in Figure 3.8 as a function of w/h, with the relative permittivity of the substrate ε_r as parameter.

In the quasi-TEM approximation with lossless conductors, the characteristic impedance does not vary with frequency. To realize low-impedance lines (with typically less than 10Ω), one requires wide strips, whereas for very large impedances ($> 150\Omega$) the strips become very narrow and may be difficult to realize.

3.4.3 Microstrip with Conductor of Finite Thickness

The actual thickness $b \neq 0$ of the upper conductor sometimes becomes of the same order of magnitude as the thickness h of the substrate and must then be taken into account. This is done, approximately, by replacing the actual width w by an effective width w_e in the calculations (Figure 3.9)

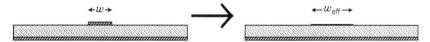

FIGURE 3.9   Definition of the equivalent strip width.

[Gunston 1972]:

$$w_e = w + \frac{b}{\pi}\left[1 + \ln\left(\frac{2x}{b}\right)\right], \tag{3.30}$$

with $x = h$ if $w > h/2\pi$ and $x = 2\pi w$ if $h/2\pi > w > 2b$.

Due to undercutting, the cross section of the conducting strip often takes a trapezoidal shape (Section 12.3.8), producing more complex effects [Railton and McGeehan 1990].

3.4.4   Phase Velocity and Line Wavelength

The phase velocity v_ϕ and the line wavelength λ_g are directly related to the effective permittivity:

$$v_\phi = c_0/\sqrt{\varepsilon_e}, \tag{3.31}$$

$$\lambda_g = \lambda_0/\sqrt{\varepsilon_e}. \tag{3.32}$$

Both the velocity and the wavelength are functions of the transmission line's geometry and, indirectly, of the characteristic impedance. The effective permittivity ε_e is shown in Figure 3.10 as a function of ε_r, with Z_c as parameter.

3.4.5   Narrow Conductor

When $w \ll h$, the fields are concentrated in the immediate vicinity of the conductor strip—that is, evenly distributed within the dielectric substrate and in the air above it. The effective permittivity is then just the average permittivity:

$$\lim_{w \ll h} \varepsilon_e = \frac{\varepsilon_r + 1}{2}. \tag{3.33}$$

The characteristic impedance then becomes

$$\lim_{w \ll h} Z_c = \frac{60\sqrt{2}}{\sqrt{\varepsilon_r + 1}} \log\left(8\frac{h}{w}\right). \tag{3.34}$$

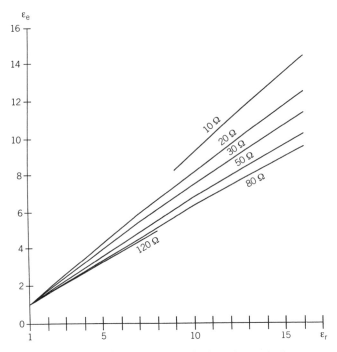

FIGURE 3.10   Effective permittivity for microstrip line.

3.4.6   Wide Conductor

At the other limit, when the strip becomes very large, most of the field is concentrated within the dielectric, so the effective permittivity is the substrate permittivity:

$$\lim_{w \gg h} \varepsilon_e = \varepsilon_r. \tag{3.35}$$

The characteristic impedance is then given by

$$\lim_{w \gg h} Z_c = \frac{120\pi}{\sqrt{\varepsilon_r}} \frac{h}{w}. \tag{3.36}$$

3.4.7   Microstrip Synthesis: The Inverse Problem

Previous sections presented the analysis of a microstrip line—the determination of the line's electrical characteristics ε_e, Z_c, and λ_g in terms of specified geometric and material parameters.

In the actual design of microstrip, one wishes to determine the width w required to obtain a specified characteristic impedance Z_c on a substrate of known permittivity ε_r and thickness h. This is the reverse operation, called synthesis.

3.4.8   Direct Synthesis Equations

Approximate expressions yielding the ratio w/h in terms of the characteristic impedance and permittivity were derived by Wheeler [1965] and touched up by Hammerstad ten years later [1975]:

when $w/h \leq 2$;

$$\frac{w}{h} \cong 4\left[\frac{1}{2}e^A - e^{-A}\right]^{-1}, \tag{3.37}$$

with

$$A = \pi\sqrt{2(\varepsilon_r + 1)}\,\frac{Z_c}{Z_0} + \frac{\varepsilon_r - 1}{\varepsilon_r + 1}\left(0.23 + \frac{0.11}{\varepsilon_r}\right),$$

where $Z_0 = 120\pi\,\Omega$; while for $w/h \geq 2$,

$$\frac{w}{h} \cong \frac{\varepsilon_r - 1}{\pi\varepsilon_r}\left[\log(B - 1) + 0.39 - \frac{0.61}{\varepsilon_r}\right] + \frac{2}{\pi}[B - 1 - \log(2B - 1)], \tag{3.38}$$

with

$$B = \frac{\pi}{2\sqrt{\varepsilon_r}}\frac{Z_0}{Z_c}.$$

These synthesis expressions, derived for an infinitely thin upper strip, are somewhat less accurate ($\leq 1\%$ error) than the ones presently available for microstrip analysis.

3.4.9   Iterative Methods, Optimization

When more accurate values are really needed, an iterative process can be established, introducing the analysis relationships within an optimization loop (Figure 3.11).

The value given by the synthesis expressions of Section 3.4.8 can be used as a starting value, and the value of w/h refined by successive approximations within the loop. It is also possible in this manner to take into account the finite thickness of the conductor strip (Section 3.4.3) and the dispersion (Section 3.5).

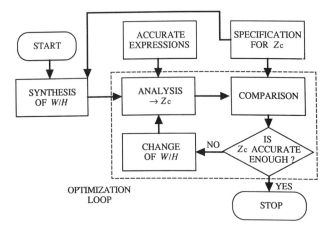

FIGURE 3.11   Schematic representation of the optimization process.

3.4.10   Effect of Cover and Walls

The previous developments considered "uncovered" lines, in which the air-filled half-space above the microstrip extends to infinity, while the ground plane and the substrate extend transversely to infinity. In practice, a circuit is generally placed in a box, with walls and covers rather close by. The characteristic impedance and the effective permittivity are affected in ways difficult to determine accurately [Hoffmann 1987]. Rules of thumb roughly indicate that for alumina ($\varepsilon_r = 9.8$), the height up to the cover should be more than eight times the substrate thickness and the distance to walls more than five times the conductor thickness for the relations to be valid.

3.5   DISPERSION IN MICROSTRIP LINES

3.5.1   Limitations of the Quasi-TEM Approximation

The quasi-TEM approximation, based on the dc distribution of the fields, yields rigorous expressions only when $f = 0$. At low frequencies, these expressions provide a good approximation. As frequency increases, the energy tends to concentrate within the dielectric. The longitudinal field components increase, and the hybrid character of the dominant mode becomes significant. The use of the quasi-TEM approximation can be somewhat extended by defining a frequency-dependent effective permittivity which increases with frequency.

At the very high frequency limit, the fields are entirely contained within the dielectric, in which case the effective permittivity becomes the relative permittivity of the dielectric.

3.5.2 Approximate Expression for ε_{ef}

Getsinger [1973] carried out a dynamic analysis of the field distribution in the microstrip, based on a simplified model of the microstrip: a section of parallel-plate dielectric-loaded waveguide located between two air-filled sections. The geometric dimensions of the equivalent model were determined experimentally. A simple formula for the effective permittivity of microstrip is then obtained:

$$\varepsilon_{ef}(f) = \varepsilon_r - \frac{\varepsilon_r - \varepsilon_{es}}{1 + G(f/f_d)^2}, \tag{3.39}$$

where ε_{es} is the low-frequency permittivity (static), defined in Section 3.4.1, ε_{ef} is the frequency-dependent permittivity, f is the signal's frequency, and the other parameters are

$$f_d = Z_c/2\mu_0 h, \tag{3.40}$$
$$G = 0.6 + 0.009 Z_c \quad [\Omega]. \tag{3.41}$$

This relation provides a simple approximation, showing in particular above which frequency this effect must be taken into account.

3.5.3 More Accurate Derivation

Kirschning and Jansen [1982] carried out a thorough analysis of the dominant hybrid mode on microstrip and derived the expression

$$\varepsilon_{ef}(f) = \varepsilon_r - \frac{\varepsilon_r - \varepsilon_{es}}{1 + P}, \tag{3.42}$$

with

$$P = P_1 P_2 \{(0.1844 + P_3 P_4) 10 fh\}^{1.5763}, \tag{3.43}$$

$$P_1 = 0.27488 + \frac{w}{h}\left\{0.6315 + \frac{0.525}{(1 + 0.157 fh)^{20}}\right\} - 0.065683 \exp\left(-8.7513\frac{w}{h}\right),$$

$$P_2 = 0.33622\{1 - \exp(-0.03442\varepsilon_r)\},$$

$$P_3 = 0.0363 \exp(-4.6w/h)\left[1 - \exp\{-(fh/3.87)^{4.97}\}\right],$$

$$P_4 = 1 + 2.751\{1 - \exp(-\varepsilon_r/15.916)^8\},$$

where the product fh is expressed in GHz · cm. The difference between the value given by Eq. 3.42 and the exact analysis results for the dominant hybrid mode is reported as being less than 0.6%.

Time

FIGURE 3.12   Widening of a Gaussian pulse due to dispersion.

3.5.4   Effect of Dispersion

The propagation factor (phase shift per unit length) is

$$\beta = \omega\sqrt{\varepsilon_0\varepsilon_{\text{ef}}\mu_0} = \frac{\omega}{c_0}\sqrt{\varepsilon_{\text{ef}}}. \tag{3.44}$$

When the permittivity varies with frequency (3.42), β is not a linear function of ω, so the phase and the group velocities are frequency dependent. Different spectral components (Section 2.5.1) of an information-carrying signal then propagate with different velocities, and the shape of the signal is distorted (phase distortion). A pulse propagating along the microstrip tends to widen as it travels. As an example, the width τ of a Gaussian pulse, defined at its $e^{-1/2}$ points, increases with distance as [Lee 1986] (Figure 3.12)

$$\sqrt{\tau^2 + \left(\frac{4z}{\tau}\frac{\partial^2\beta}{\partial\omega^2}\right)^2}. \tag{3.45}$$

In practical applications, a sufficient free time should be kept between successive pulses so that they can still be properly discriminated when reaching the receiver.

3.5.5   Recommendation

The dispersion is directly related to the product fh, which should be kept as small as possible to prevent pulse broadening. Dispersion also modifies the characteristic impedance and introduces uncertainty (Section 3.3.3). The substrate should therefore be thinned as the signal frequency increases.

3.5.6   Remark

The relative permittivity of the substrate also varies with frequency when the substrate is lossy (Section 2.4.4). This effect is small with low-loss dielectrics, but can become significant for semiconductor substrates.

3.6   LOSSES IN MICROSTRIP

3.6.1   Definitions

Losses are a fact of life: no matter how carefully a circuit is designed, a part of the signal is transformed into useless heat and lost as far as communication is concerned. Given sufficient time, any response eventually dies out. The amplitude of the signal decreases as it travels along the line; the signal is attenuated.

If attenuation were the same for all signals, the original signal could simply be restored by amplification. However, losses are frequency dependent, producing an additional source of dispersion, which creates amplitude distortion.

Losses result from two mechanisms: conduction and dissipation in dielectric and magnetic materials. While the end results are basically similar, the electromagnetic analysis is strongly dependent upon the location of the losses.

3.6.2   Propagation on Lossy Transmission Line

The propagation factor on a lossy transmission line is complex:

$$\underline{\gamma} = \alpha + j\beta. \tag{3.46}$$

The real part α is the attenuation per unit length, while the imaginary part β is the phase shift per unit length. The attenuation $\alpha = \alpha_d + \alpha_m$ results from substrate losses (α_d, Section 3.6.3) and conductor losses (α_m, Section 3.6.4). The voltage and current on a lossy microstrip line are then

$$\underline{U}(z) = \underset{\text{forward wave}}{\underline{U}_+ \exp(-\underline{\gamma}z)} + \underset{\text{reverse wave}}{\underline{U}_- \exp(+\underline{\gamma}z)},$$

$$\underline{I}(z) = \underset{\text{forward wave}}{\underline{Y}_c\underline{U}_+ \exp(-\underline{\gamma}z)} - \underset{\text{reverse wave}}{\underline{Y}_c\underline{U}_- \exp(+\underline{\gamma}z)}. \tag{3.47}$$

The losses in the equivalent circuit of a transmission line are represented by resistive terms in the line model (Figure 3.13). Losses within the substrate contribute a shunt conductance, and metal losses a series resistance.

3.6.3   Nepers and Decibels

The linear attenuation α that appears in Eq. 3.46 is expressed in Nepers per unit length. Their use is directly related to base e exponentials, the amplitude of a forward wave being then given by

$$|\underline{U}(z)| = |\underline{U}(0)| \exp(-\alpha_{Np}z). \tag{3.48}$$

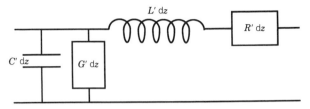

FIGURE 3.13   Equivalent circuit for a length dz of a lossy transmission line.

In actual practice, one prefers to use a base 10 formulation, as this considerably simplifies calculations. Attenuation is then defined in decibels (dB) per unit length, with the following amplitude dependence for a forward wave:

$$|\underline{U}(z)| = |\underline{U}(0)|10^{-\alpha_{dB}z/20}. \tag{3.49}$$

The two attenuations are then related by

$$\alpha_{dB} = (20/\ln 10)\alpha_{Np} \cong 8.686\,\alpha_{Np}. \tag{3.50}$$

Decibel and Neper values for some amplitude ratios are given in Table 3.1.

In the following sections, linear attenuation values are given in decibels per meter (dB/m).

3.6.4   Substrate Losses

Losses within the dielectric substrate do not modify the basic electromagnetic field distribution. Within an insulating material, they are represented by the imaginary part of the permittivity ε''. For a line deposited on a semiconductor substrate, conduction losses also play a significant role and result from the presence of a substrate conductivity σ. As far as transmission on the line is concerned, the two loss mechanisms are equivalent. To simplify developments, all losses within the propagation medium will be included within the permittivity term; hence, ε'' contains both dielectric and conduction losses ($\varepsilon'' + \sigma/\omega$ is replaced by ε'' alone).

Magnetic losses play a role only for ferrite substrates. Other substrates are nonmagnetic, with real permeability μ_0.

The attenuation produced by the microstrip substrate can be determined by [Hammerstad and Bekkadal 1975]

$$\alpha_d \cong 27.3\frac{\varepsilon_{ef} - 1}{\varepsilon_r - 1}\frac{\varepsilon_r}{\varepsilon_{ef}}\frac{\tan\delta}{\lambda_g} \quad [\text{dB/m}], \tag{3.51}$$

TABLE 3.1 Amplitude Ratios in Np and dB

Decibels	Nepers	Voltage Ratio	Power Ratio
0.1	0.0115	0.9886	0.9772
0.2	0.0230	0.9772	0.9550
0.3	0.0345	0.9661	0.9333
0.4	0.0461	0.9550	0.9120
0.5	0.0576	0.9441	0.8913
0.8	0.0921	0.9120	0.8317
1.0	0.1151	0.8913	0.7943
1.5	0.1727	0.84134	0.7080
2.0	0.2303	0.7943	0.6310
3.0	0.3454	0.7079	0.5012
5.0	0.5756	0.5623	0.3162
8.0	0.9210	0.3981	0.1584
10.0	1.1513	0.3162	0.1
15.0	1.7269	0.1778	0.0316
20.0	2.3026	0.1	0.01
30.0	3.4538	0.0316	0.001
40.0	4.6051	0.01	0.0001
100.0	11.5128	0.00001	0.0000000001
0.8686	0.1	0.9048	0.8187
1.7372	0.2	0.8187	0.6703
2.6058	0.3	0.7408	0.5488
3.4744	0.4	0.6703	0.4493
4.3430	0.5	0.6065	0.3678
6.9488	0.8	0.4493	0.2018
8.6860	1.0	0.3679	0.1353
13.0290	1.5	0.2231	0.04978
17.3720	2.0	0.1353	0.018314
26.0580	3.0	0.0497	0.002479
34.7440	4.0	0.0183	0.0003354
43.4300	5.0	0.00674	0.00004539
69.4880	8.0	0.000335	0.000000112
86.8600	10.0	0.000045	0.0000000021
40.0000	4.6051	0.01	0.0001
33.9794	3.9120	0.02	0.0004
30.4576	3.5065	0.03	0.0009
27.9588	3.2188	0.04	0.0016
26.0206	2.9957	0.05	0.0025
21.9382	2.5257	0.08	0.0064
20.0000	2.3026	0.1	0.01
16.4782	1.8971	0.15	0.0225
13.9794	1.6094	0.2	0.04
10.4576	1.2040	0.3	0.09
7.9588	0.9163	0.4	0.16
6.0206	0.6931	0.5	0.25
1.9382	0.2231	0.8	0.64
0.0000	0.0000	1.0	1.0

TABLE 3.1   Amplitude Ratios in Np and dB (Continued)

Decibels	Nepers	Voltage Ratio	Power Ratio
20.0000	2.3026	0.1	0.01
16.9897	1.9560	0.1414	0.02
15.2288	1.7533	0.1732	0.03
13.9794	1.6094	0.2	0.04
13.0103	1.4978	0.2236	0.05
10.9691	1.2628	0.2828	0.08
10.0000	1.1513	0.3162	0.1
8.2391	0.9485	0.3872	0.15
6.9897	0.8047	0.4472	0.2
5.2288	0.6020	0.5477	0.3
3.9794	0.4581	0.6324	0.4
3.0103	0.3466	0.7071	0.5
0.9691	0.1116	0.8944	0.8
0.0000	0.0000	1.0	1.0

where the frequency-dependent effective permittivity ε_{ef} and the guide wavelength λ_g are defined in Eqs. 3.42, and 3.32.

3.6.5   Conductor Losses

When the conductors have a finite conductivity ($\sigma \neq \infty$, Section 2.3.6), a rigorous analysis should determine the fields within the conductors, in addition to those within the substrate. The continuity of the tangential components of $\underline{\mathbf{E}}$ and $\underline{\mathbf{H}}$ provides eight equations to be satisfied at dielectric–metal interfaces. This would become extremely complex and practically impossible to carry out.

When one structure is only slightly different from another one, the difference between the two, called perturbation, can be evaluated with good accuracy [Morse and Feshbach 1953]. Comparing a microstrip with real metal conductors to one having PEC conductors, perturbation theory yields the general expression [Gardiol 1984]

$$\underline{\gamma} + \underline{\gamma}_0^* = \alpha + j(\beta - \beta_0)$$

$$\cong (1 + j)\,R_m \frac{\oint_C |\underline{H}_0|^2 \, dl}{2 \int_S \mathrm{Re}\left[\mathbf{e}_z \cdot (\underline{\mathbf{E}}_0 \times \underline{\mathbf{H}}_0^*)\right] dA} = \kappa\sqrt{j\omega}\,, \quad (3.52)$$

where R_m is the real part of the complex metal impedance defined in Section 2.6.5:

$$\underline{Z}_m = \sqrt{\frac{j\omega\mu}{\sigma}} = (1 + j)\sqrt{\frac{\omega\mu}{2\sigma}} = (1 + j)\,R_m. \quad (2.43)$$

The subscript 0 refers to the unperturbed structure. The integral in the numerator must be evaluated on all conductors, the one in the denominator over the microstrip cross section that covers the entire upper half-space.

Metal losses also affect the linear phase shift β. Although small, this contribution should not be neglected when evaluating the time-domain response, since this would yield noncausal solutions [Arabi et al. 1991].

A simple expression for the attenuation produced by metal losses was derived by Janssen [1977] and found to be accurate enough in many practical situations:

$$\alpha_{\mathrm{m}} = 8.686 R_{\mathrm{m}}/wZ_{\mathrm{c}} \quad [\mathrm{dB/m}]. \tag{3.53}$$

A more thorough analysis of the losses was carried out by Pucel et al. [1968], who obtained

$$\alpha_{\mathrm{m}} \cong \frac{1}{h} \frac{10}{\pi \ln 10} \frac{R_{\mathrm{m}}}{Z_{\mathrm{c}}} \frac{32 - (w/h)^2}{32 + (w/h)^2} \left[1 + \frac{h}{w} \left(1 + \frac{\partial w_{\mathrm{e}}}{\partial b} \right) \right] \quad [\mathrm{dB/m}]$$

$$\text{for } w \leq h,$$

$$\alpha_{\mathrm{m}} \cong \frac{1}{h} \frac{20}{\ln 10} \frac{Z_{\mathrm{c}} R_{\mathrm{m}} \varepsilon_{\mathrm{e}}}{Z_0^2} \left\{ \frac{w}{h} + \frac{6h}{w} \left[\left(1 - \frac{h}{w} \right)^5 + 0.08 \right] \right\} \left[1 + \frac{h}{w} \left(1 + \frac{\partial w_{\mathrm{e}}}{\partial b} \right) \right] \quad [\mathrm{dB/m}]$$

$$\text{for } w \geq h, \quad (3.54)$$

where $\partial w_{\mathrm{e}}/\partial b$ is determined from Eq. 3.29:

$$\frac{\partial w_{\mathrm{e}}}{\partial b} = \begin{cases} \dfrac{1}{\pi} \ln \dfrac{2h}{b} & \text{if } 2\pi w \geq h, \\[2ex] \dfrac{1}{\pi} \ln \dfrac{4\pi w}{b} & \text{if } 2\pi w \leq h. \end{cases} \tag{3.55}$$

The dimensions of the microstrip are defined in Figure 3.5. These expressions were obtained for ideal planar metallic surfaces. The attenuation α_{m} produced by metal losses is frequency dependent, since the metal resistance R_{m} in Eqs. 3.53 and 3.54 varies with $\sqrt{\omega}$ (Eq. 2.43). The characteristics of lossy microstrip can be evaluated with a calculator program [Kajfez and Tew 1980].

The range of validity of perturbation techniques is difficult to evaluate, so the method must be used carefully. It is adequate to account for the attenuation produced by conductor losses, but may no longer be accurate enough to evaluate losses produced by absorbing materials (for instance in deposited resistors, Section 12.4.2).

3.6.6 Time-Domain Response

The substrate attenuation α_d (Eq. 3.47) is approximately independent of frequency, at least in the nondispersive part of the characteristic (Figure 3.6). The finite conductivity of the metal leads to a term $\alpha_m + j\beta_m = \kappa\sqrt{j\omega}$ in which the value of κ can be extracted from Eq. 3.51:

$$\kappa \cong \frac{1}{wZ_c}\sqrt{\frac{\mu}{2\sigma}}. \tag{3.56}$$

The response to a pulse of length τ_0 Eq. 2.25 is given by [Gardiol 1987]

$$u(z,t) = \frac{1}{2\pi}\int_{-\infty}^{\infty}\frac{1}{j\omega}\exp\left[j\omega\left(t - \frac{z}{c}\right) - \alpha_d z - \kappa z\sqrt{j\omega}\right]d\omega$$

$$-\frac{1}{2\pi}\int_{-\infty}^{\infty}\frac{1}{j\omega}\exp\left[j\omega\left(t - \tau_0 - \frac{z}{c}\right) - \alpha_d z - \kappa z\sqrt{j\omega}\right]d\omega$$

$$= \exp(-\alpha_d z)\,\text{erfc}\,\frac{\kappa z}{2\sqrt{t - z/c}}\mathbf{1}\left(t - \frac{z}{c}\right)$$

$$- \exp(-\alpha_d z)\,\text{erfc}\,\frac{\kappa z}{2\sqrt{t - \tau_0 - z/c}}\mathbf{1}\left(t - \tau_0 - \frac{z}{c}\right), \tag{3.57}$$

where

$$\mathbf{1}(t) = \begin{cases} 0 & \text{for } t < 0, \\ 1 & \text{for } t > 0. \end{cases}$$

The dielectric attenuation α_d must be expressed in Np/m (Section 3.6.3). It produces an exponential decrease in amplitude as the pulse travels along the line, whereas the pulse is distorted by the frequency-dependent metal attenuation (Figure 3.14).

A more complete derivation, also accounting for the frequency-dependent effective permittivity ε_{ef} and the frequency variation of ε_r, would no longer yield an analytical expression like Eq. 3.57, but the inverse Fourier transform should be determined by a computer technique such as FFT.

3.6.7 Remark

In most microstrip circuits the connecting lines are rather short, and the effects of losses and dispersion remain small. When longer sections of microstrip are used, as in delay lines, filters, and feed networks for antennas, these effects become significant and must be considered.

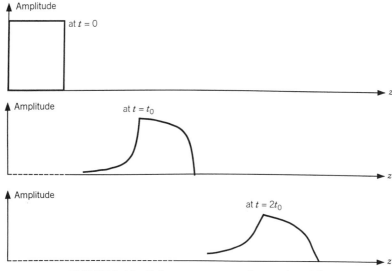

FIGURE 3.14   Pulse response on a lossy microstrip.

3.7   FREQUENCY LIMITATIONS

The frequency band over which a microstrip can be used is limited by the spurious effects that appear when frequency increases. Recommendations as to the values of the *fh* product that should not be exceeded are given here, whereas the techniques used to analyze high-frequency effects are provided in Chapter 10.

3.7.1   Higher-Order Modes

At a low frequency, the sole dominant mode propagates along the microstrip line (Section 3.1.3, Figure 3.7). When the signal frequency exceeds the cutoff of the first higher-order mode, two modes can propagate with different velocities. The signal transmitted to the end of the line is then a combination of the two modes, which is distorted because the two components are not in phase. To avoid this effect, one operates below the cutoff frequency of the first higher-order mode, which is approximately given by Vendelin [1970] as

$$f_c \cong \frac{c_0}{\sqrt{\varepsilon_r}\,(2w + 0.8h)}. \tag{3.58}$$

It is the cutoff frequency of the dominant mode in a waveguide having an equivalent width of $2w + 0.8h$, with c.m.p. boundaries on both sides. This limitation depends mostly on the strip width and will primarily affect wide strips (i.e., low-impedance lines).

3.7.2 Surface Waves

Surface waves are the propagating modes of the planar metal–dielectric–air structure (microstrip without upper conductor), with fields mostly trapped within the dielectric substrate (Figure 1.5, Section 1.3.5). The lowest-order TM surface-wave mode can propagate at any frequency (it has no cutoff). However, its coupling to the quasi-TEM mode of microstrip only becomes significant when the phase velocities of the two modes are close [Vendelin 1970]. Synchronism occurs at the frequency

$$f_s = \frac{c_0 \tan^{-1} \varepsilon_r}{\sqrt{2}\, \pi h \sqrt{\varepsilon_r - 1}}. \tag{3.59}$$

3.7.3 Radiation

Since the microstrip line is an open structure (i.e., not enclosed within a metal envelope), some higher-order modes are radiating modes. An infinite straight transmission line propagating the dominant mode does not radiate: there is no coupling between the fields of the dominant mode and those of radiated modes. However, as soon as some discontinuity appears along the line, higher-order radiating modes are excited. An approximate value for the frequency at which the radiation becomes significant was determined from data published by Hammerstad and Bekkadal [1975]:

$$f\,[\text{GHz}]\, h\,[\text{mm}] > 2.14 \sqrt[4]{\varepsilon_r}. \tag{3.60}$$

3.7.4 Comment

Three different mechanisms may limit the operation of microstrip lines. It is therefore recommended to calculate the three limiting frequencies and then take the worst case (i.e., the lowest value).

3.8 PROBLEMS

We wish to build a microstrip line with a characteristic impedance of 80 Ω on a polypropylene substrate that has the following characteristics:

relative permittivity $\varepsilon_r' = 2.18 \pm 0.05$

thickness $h = 0.8 \pm 0.04$ mm

3.8.1   Determine the width w of the upper conductor, making use of Wheeler's relationships (Section 3.4.8). At this point we assume that the conductor thickness can be neglected and that the substrate parameters are at their nominal value (± 0).

3.8.2   Using the analysis equations (Sections 3.4.1 and 3.4.2) or similar ones given in other books or CAD softwares, determine the effective permittivity ε_e and the characteristic impedance Z_c: calculate the error with respect to the 80 Ω requested.

3.8.3   Determine the maximum changes in the line parameters for substrate properties at the limits of the specification range.

3.8.4   What effects would an error of $\pm 10\%$ in the width of the upper conductor produce?

3.8.5   Determine the correction of width required to take into account the thickness of the upper conductor $b = 50\ \mu$m.

3.8.6   Determine the attenuation due to the substrate losses, knowing that $\tan \delta = 0.0003$ for polypropylene at 10 GHz.

3.8.7   What are the conductor losses for copper at 1, 3, 5, and 10 GHz?

3.8.8   Evaluate approximately the effect of dispersion (Sections 3.5.2 and 3.5.3) at a signal frequency of 12 GHz. How can one compensate for it? At which frequency does it become significant?

3.8.9   Find out at which frequency the microstrip line starts radiating.

3.8.10   Determine the size reduction that is obtained for the longitudinal and transverse dimensions when the polypropylene substrate is replaced by a substrate having a relative permittivity of 80 for the same line characteristics.

CHAPTER FOUR

Devices

A printed circuit is made by assembling different devices and connecting them by sections of microstrip lines. Each device is characterized by an equivalent circuit or a matrix. The behavior of the complete circuit is determined by the combination of all the circuit components.

4.1 GENERAL DESCRIPTION

4.1.1 Definition

The term *microstrip device* defines a structure connected to n microstrip lines (Figure 4.1). The dimensions of the connecting lines are selected so that only one mode, the dominant mode, can propagate.

4.1.2 Higher-Order Modes

At the interface between a connecting line and the device, the boundary conditions on the line side cannot be satisfied by the field components of the dominant mode alone. Several modes are necessary to ensure the continuity of the current so that higher-order modes appear in the close vicinity of the discontinuity.

4.1.3 Evanescent Guided Modes

The line dimensions were selected in such a way that only one mode can propagate, meaning that higher-order guided modes are evanescent. The field amplitudes of these modes taper off quickly as one moves away along the lines. Evanescent modes cannot transport power, except when another discontinuity is very close to the first one. An evanescent mode storing

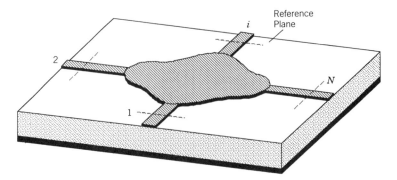

FIGURE 4.1 General description of an n-port device.

predominantly electric energy produces a capacitive effect, whereas a mode that mostly stores magnetic energy gives rise to an inductive effect.

4.1.4 Radiated and Surface-Wave Modes

Since microstrips are open transmission lines, radiated and surface waves can also be excited at the edge of a circuit. In this case, some part of the signal vanishes, and the resulting loss of power can be represented by resistive components in the description of the device.

Waves excited at a discontinuity can also be picked up by other parts of the circuit, producing spurious coupling. This effect involves several devices, so it is nonlocal. Its analysis becomes rather complicated, since it requires the introduction of controlled signal sources at sensitive points within the circuit. This effect becomes significant when the frequency of the signal is large. Its onset provides a practical limit to the useful band of the structure.

4.1.5 Reference Plane

On every access line i connecting a device to the outer world, a coordinate axis z_i is defined. The origin of the coordinates ($z_i = 0$) is the reference plane at port i. This plane must be located far enough from the device so that the effect of the evanescent modes can be neglected. The access lines are assumed to be lossless.

4.2 IMPEDANCE, ADMITTANCE, AND CHAIN MATRICES

4.2.1 Currents and Voltages

A microstrip device can be characterized in terms of the currents entering its n ports and of the voltages across the conductors at its n ports (Figure 4.2)

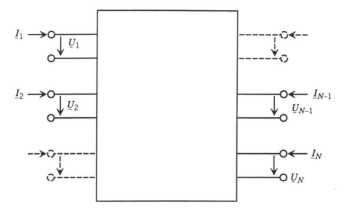

FIGURE 4.2   Currents and voltages at the ports of a device.

4.2.2 Impedance Matrix

The impedance matrix of an n-port device links the voltages to the currents at its n ports and is defined by

$$
\begin{pmatrix} \underline{U}_1 \\ \underline{U}_2 \\ \cdots \\ \underline{U}_i \\ \cdots \\ \underline{U}_n \end{pmatrix}
=
\begin{pmatrix}
\underline{Z}_{11} & \underline{Z}_{12} & \cdots & \underline{Z}_{1i} & \cdots & \underline{Z}_{1n} \\
\underline{Z}_{21} & \underline{Z}_{22} & \cdots & \underline{Z}_{2i} & \cdots & \underline{Z}_{2n} \\
\cdots & \cdots & \cdots & \cdots & \cdots & \cdots \\
\underline{Z}_{i1} & \underline{Z}_{i2} & \cdots & \underline{Z}_{ii} & \cdots & \underline{Z}_{in} \\
\cdots & \cdots & \cdots & \cdots & \cdots & \cdots \\
\underline{Z}_{n1} & \underline{Z}_{n2} & \cdots & \underline{Z}_{ni} & \cdots & \underline{Z}_{nn}
\end{pmatrix}
\begin{pmatrix} \underline{I}_1 \\ \underline{I}_2 \\ \cdots \\ \underline{I}_i \\ \cdots \\ \underline{I}_n \end{pmatrix} . \quad (4.1)
$$

The term $\underline{Z}_{ij}$ is the ratio of the voltage $\underline{U}_i$ to the current $\underline{I}_j$ when the currents vanish at all other ports ($\underline{I}_k = 0$ for $k \neq j$)—that is, with all ports but one open-circuited. The term $\underline{Z}_{ii}$ is the input impedance to the device at port i when no current is injected into any output port.

4.2.3 Admittance Matrix

In a similar manner, the admittance matrix of an n-port device links the currents to the voltages at its n ports and is defined by

$$
\begin{pmatrix} \underline{I}_1 \\ \underline{I}_2 \\ \cdots \\ \underline{I}_i \\ \cdots \\ \underline{I}_n \end{pmatrix}
=
\begin{pmatrix}
\underline{Y}_{11} & \underline{Y}_{12} & \cdots & \underline{Y}_{1i} & \cdots & \underline{Y}_{1n} \\
\underline{Y}_{21} & \underline{Y}_{22} & \cdots & \underline{Y}_{2i} & \cdots & \underline{Y}_{2n} \\
\cdots & \cdots & \cdots & \cdots & \cdots & \cdots \\
\underline{Y}_{i1} & \underline{Y}_{i2} & \cdots & \underline{Y}_{ii} & \cdots & \underline{Y}_{in} \\
\cdots & \cdots & \cdots & \cdots & \cdots & \cdots \\
\underline{Y}_{n1} & \underline{Y}_{n2} & \cdots & \underline{Y}_{ni} & \cdots & \underline{Y}_{nn}
\end{pmatrix}
\begin{pmatrix} \underline{U}_1 \\ \underline{U}_2 \\ \cdots \\ \underline{U}_i \\ \cdots \\ \underline{U}_n \end{pmatrix} . \quad (4.2)
$$

The term $\underline{Y}_{ij}$ is the ratio of the current $\underline{I}_i$ to the voltage $\underline{U}_j$ when the voltages vanish at all other ports ($\underline{U}_k = 0$ for $k \neq j$)—that is, when all ports but one are short-circuited. The term $\underline{Y}_{ii}$ is the input admittance to the device at port i when no voltage is applied at output ports.

4.2.4 Relationship between Impedance and Admittance Matrices

The admittance and the impedance matrices are inverses of each other (if the respective determinant is nonzero):

$$(\underline{Y}) = (\underline{Z})^{-1} \quad \text{and} \quad (\underline{Z}) = (\underline{Y})^{-1}. \tag{4.3}$$

In the case of a two-port, one gets

$$Y_{11} = \frac{\underline{Z}_{22}}{\underline{Z}_{11}\underline{Z}_{22} - \underline{Z}_{12}\underline{Z}_{21}}, \quad Y_{12} = \frac{-\underline{Z}_{12}}{\underline{Z}_{11}\underline{Z}_{22} - \underline{Z}_{12}\underline{Z}_{21}},$$

$$Y_{21} = \frac{-\underline{Z}_{21}}{\underline{Z}_{11}\underline{Z}_{22} - \underline{Z}_{12}\underline{Z}_{21}}, \quad Y_{22} = \frac{\underline{Z}_{11}}{\underline{Z}_{11}\underline{Z}_{22} - \underline{Z}_{12}\underline{Z}_{21}}, \tag{4.4}$$

and

$$\underline{Z}_{11} = \frac{Y_{22}}{Y_{11}Y_{22} - Y_{12}Y_{21}}, \quad \underline{Z}_{12} = \frac{-Y_{12}}{Y_{11}Y_{22} - Y_{12}Y_{21}},$$

$$\underline{Z}_{21} = \frac{-Y_{21}}{Y_{11}Y_{22} - Y_{12}Y_{21}}, \quad \underline{Z}_{21} = \frac{Y_{11}}{Y_{11}Y_{22} - Y_{12}Y_{21}}. \tag{4.5}$$

4.2.5 Chain or *ABCD* Matrix of a Two-Port

In many applications, one wishes to know the parameters at the output (port 2) of a two-port device in terms of those at its input (port 1). The relationship is expressed by a 2×2 matrix, called the *ABCD* or chain matrix:

$$\begin{pmatrix} U_2 \\ -I_2 \end{pmatrix} = \begin{pmatrix} A & B \\ C & D \end{pmatrix} \begin{pmatrix} U_1 \\ I_1 \end{pmatrix}. \tag{4.6}$$

The *ABCD* coefficients can be determined from the impedance or admittance matrices:

$$A = \frac{\underline{Z}_{22}}{\underline{Z}_{12}} = -\frac{Y_{11}}{Y_{12}}, \quad B = \frac{\underline{Z}_{12}\underline{Z}_{21} - \underline{Z}_{11}\underline{Z}_{22}}{\underline{Z}_{12}} = \frac{1}{Y_{12}},$$

$$C = \frac{-1}{\underline{Z}_{12}} = \frac{Y_{11}Y_{22} - Y_{12}Y_{21}}{Y_{12}}, \quad D = \frac{\underline{Z}_{11}}{\underline{Z}_{12}} = -\frac{Y_{22}}{Y_{12}}. \tag{4.7}$$

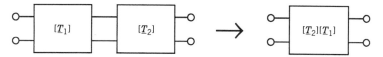

FIGURE 4.3   Cascading of a chain of two-port devices.

The *ABCD* matrices present a particular interest for cascading a chain of two-port devices (Figure 4.3) because the *ABCD* matrix of the cascade connection is obtained by simply multiplying the matrices of the individual devices.

4.3   SCATTERING MATRIX

4.3.1   Remark

At very high frequencies and at microwaves, it is not desirable to set up open- or short-circuit conditions because they might give rise to resonances. Also, the concepts of current and voltage are not uniquely defined in non-TEM lines (more particularly so in the case of waveguides). For these reasons, the concept of the scattering matrix (or of *S*-parameters) was introduced [Montgomery et al. 1948].

4.3.2   Complex Normalized Waves

At port i of the device, the complex normalized waves $\underline{a}_i$ and $\underline{b}_i$ are defined by

$$\underline{a}_i = \frac{\underline{U}_i + Z_{ci}\underline{I}_i}{2\sqrt{Z_{ci}}} \qquad \text{and} \qquad \underline{b}_i = \frac{\underline{U}_i - Z_{ci}\underline{I}_i}{2\sqrt{Z_{ci}}}. \tag{4.8}$$

where Z_{ci} is the characteristic impedance of the transmission line at port i.

The dimensions of $\underline{a}_i$ and $\underline{b}_i$ are the square root of power [$W^{1/2}$]. The current and voltage at port i are then

$$\underline{U}_i = \sqrt{Z_{ci}}(\underline{a}_i + \underline{b}_i) \qquad \text{and} \qquad \underline{I}_i = (\underline{a}_i - \underline{b}_i)/\sqrt{Z_{ci}}. \tag{4.9}$$

Introducing the values of the current and of the voltage on the line, given by Eq. 3.26, yields

$$\underline{a}_i = \left(\underline{U}_{i+}/\sqrt{Z_{ci}}\right)\exp(-j\beta_i z_i) \qquad \text{and} \qquad \underline{b}_i = \left(\underline{U}_{i-}/\sqrt{Z_{ci}}\right)\exp(+j\beta_i z_i). \tag{4.10}$$

It is thus apparent that $\underline{a}_i$ represents the signal that enters the device at port i, while $\underline{b}_i$ corresponds to the outgoing signal (reflected or transmitted from another port). The active power at port i of the device is defined by

$$P_i = \text{Re}(\underline{U}_i\underline{I}_i^*) = \text{Re}\left[(\underline{a}_i + \underline{b}_i)(\underline{a}_i - \underline{b}_i)^*\right] = |\underline{a}_i|^2 - |\underline{b}_i|^2. \quad (4.11)$$

The total active power is thus the difference between the power of the signal entering the device $|\underline{a}_i|^2$ and the power of the signal leaving the device $|\underline{b}_i|^2$.

4.3.3   Scattering Matrix

Incoming and outgoing waves are related to each other by algebraic relations. One then has a set of n equations with n unknowns that can be expressed in matrix form:

$$\begin{pmatrix} \underline{b}_1 \\ \underline{b}_2 \\ \cdots \\ \underline{b}_i \\ \cdots \\ \underline{b}_n \end{pmatrix} = \begin{pmatrix} \underline{s}_{11} & \underline{s}_{12} & \cdots & \underline{s}_{1i} & \cdots & \underline{s}_{1n} \\ \underline{s}_{21} & \underline{s}_{22} & \cdots & \underline{s}_{2i} & \cdots & \underline{s}_{2n} \\ \cdots & \cdots & \cdots & \cdots & \cdots & \cdots \\ \underline{s}_{i1} & \underline{s}_{i2} & \cdots & \underline{s}_{ii} & \cdots & \underline{s}_{in} \\ \cdots & \cdots & \cdots & \cdots & \cdots & \cdots \\ \underline{s}_{n1} & \underline{s}_{n2} & \cdots & \underline{s}_{ni} & \cdots & \underline{s}_{nn} \end{pmatrix} \begin{pmatrix} \underline{a}_1 \\ \underline{a}_2 \\ \cdots \\ \underline{a}_i \\ \cdots \\ \underline{a}_n \end{pmatrix}. \quad (4.12)$$

This expression can be written in an abbreviated form as

$$(\underline{b}) = (\underline{s})(\underline{a}). \quad (4.13)$$

The matrix $(\underline{s})$ is called the *scattering matrix*.

4.3.4   Transform Functions

Each $\underline{s}_{ij}$ parameter in the scattering matrix is a transfer function, given by the ratio of the output signal $\underline{b}_i$ at port i to the input signal $\underline{a}_j$ at port j when no signal is fed at any other port $k \neq j$ (i.e., when all these ports are terminated into nonreflecting matched loads).

When $i = j$, the term $\underline{s}_{ii}$ on the main diagonal of the scattering matrix is the *intrinsic reflection factor* of the device; that is, it represents the reflection at the input i when all the outputs are matched.

When $i \neq j$, the term is off-diagonal and is the *transmission factor* from port j to port i of the device when all outputs are matched. It must be clearly noted that the term $\underline{s}_{ij}$ corresponds to the transfer from port j to port i, *not from port i to port j.*

4.3.5   Signal Flow Graphs

The physical meaning of the terms in the scattering matrix can be illustrated by means of flow graphs. Every port is represented by two nodes, one at which the incoming signal arrives, the other from which the outgoing signal

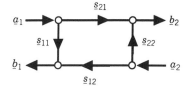

FIGURE 4.4 Flow graph of a two-port device.

departs. Each term $\underline{s}_{ij}$ is associated with one arrow, pointing from input node j to output node i. The signal flow graph of a two-port device is shown in Figure 4.4.

The value taken by the signal at any given node is obtained by summing the contributions of all the arrows reaching the node. A port arrow designates an incoming or outgoing signal. The contribution of a device arrow is the product of the transfer function associated with the arrow by the value of the signal at the point from which the arrow originates. Signal flow graphs are of particular interest for studying the interconnections of several devices. An example is shown in Figure 4.5 for the cascading of two two-port devices.

The scattering matrix of this assembly is given by

$$\underline{s}_{11}^{C} = \underline{s}_{11}^{A} + \frac{\underline{s}_{21}^{A}\underline{s}_{11}^{B}\underline{s}_{12}^{A}}{1 - \underline{s}_{11}^{B}\underline{s}_{22}^{A}}, \qquad \underline{s}_{12}^{C} = \frac{\underline{s}_{12}^{B}\underline{s}_{12}^{A}}{1 - \underline{s}_{11}^{B}\underline{s}_{22}^{A}},$$

$$\underline{s}_{21}^{C} = \frac{\underline{s}_{21}^{A}\underline{s}_{21}^{B}}{1 - \underline{s}_{11}^{B}\underline{s}_{22}^{A}}, \qquad \underline{s}_{22}^{C} = \underline{s}_{22}^{B} + \frac{\underline{s}_{12}^{B}\underline{s}_{22}^{A}\underline{s}_{21}^{B}}{1 - \underline{s}_{11}^{B}\underline{s}_{22}^{A}}, \qquad (4.14)$$

where the superscript A refers to the left-hand two-port, B to the right-hand one, and C is related to the cascade connection. The presence of a loop between the two devices leads to the apparition in the denominator of the term $1 -$ loop factor.

4.3.6 Translation of the Reference Planes

Shifting the reference plane at port i by a distance Δz_i toward the device produces a phase shift of all the terms of the scattering matrix related to port

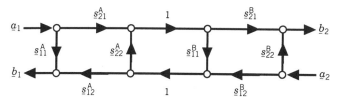

FIGURE 4.5 Flow graph of two cascaded two-port devices.

i. The shifted term $\underline{s}_{ij}^{sh}$ takes the form [Gardiol 1984]

$$\underline{s}_{ij}^{sh} = \underline{s}_{ij} \exp\left[-j(\beta_i \Delta z_i + \beta_j \Delta z_j)\right]. \qquad (4.15)$$

Some demonstrations can be simplified by selecting reference planes in such a way that some terms become purely real or purely imaginary.

4.3.7 Matrix Transformations

It is possible to derive the scattering matrix from the impedance or the admittance matrices by a set of matrix relations:

$$(\underline{s}) = (F)[(\underline{Z}) - (G)][(\underline{Z}) + (G)]^{-1}(F)^{-1},$$

$$(\underline{s}) = (F)[(1) - (G)(\underline{Y})][(1) + (G)(\underline{Y})]^{-1}(F)^{-1}, \qquad (4.16)$$

where (1) is the unit $n \times n$ matrix, and where the diagonal matrices (F) and (G) are defined by

$$(F) = \left(\text{diag}\,\frac{1}{2\sqrt{Z_{ci}}}\right) = \begin{pmatrix} \dfrac{1}{2\sqrt{Z_{c1}}} & 0 & \cdots & 0 \\ 0 & \dfrac{1}{2\sqrt{Z_{c2}}} & \cdots & 0 \\ \cdots & \cdots & \cdots & \cdots \\ 0 & 0 & \cdots & \dfrac{1}{2\sqrt{Z_{cn}}} \end{pmatrix}, \qquad (4.17)$$

$$(G) = (\text{diag}\,Z_{ci}) = \begin{pmatrix} Z_{c1} & 0 & \cdots & 0 \\ 0 & Z_{c2} & \cdots & 0 \\ \cdots & \cdots & \cdots & \cdots \\ 0 & 0 & \cdots & Z_{cn} \end{pmatrix}. \qquad (4.18)$$

The impedance and the admittance matrices can similarly be obtained in terms of the scattering matrix:

$$(\underline{Z}) = (F)^{-1}[(1) + (\underline{s})][(1) - (\underline{s})]^{-1}(F)(G),$$

$$(\underline{Y}) = (G)^{-1}(F)^{-1}[(1) - (\underline{s})][(1) + (\underline{s})]^{-1}(F). \qquad (4.19)$$

4.3.8 Input VSWR

The intrinsic reflection characteristic of a device at its port i is often represented by its voltage standing-wave radio (VSWR), which is defined by the ratio of maximum to minimum voltages on the connecting line. All the other ports of the device are terminated by reflectionless matched loads (Section 4.4.5).

$$\text{VSWR}_i = \frac{1 + |\underline{s}_{ii}|}{1 - |\underline{s}_{ii}|}. \tag{4.20}$$

This ratio is always larger than unity. A small reflection, corresponding to a well-matched device, gives a VSWR value close to 1. Larger reflections provide larger values for VSWR, with total reflection (short circuit, open circuit, purely reactive loads) corresponding to infinity.

The input reflection is also defined by the reflected power ratio in decibels (dB):

$$LR_i = -10 \log_{10} \frac{P_r}{P_{\text{in}}} = -20 \log_{10} |\underline{s}_{ii}| \quad [\text{dB}], \tag{4.21}$$

where P_r is the reflected power and P_{in} is the incident power. Decibels are the units of attenuation (Section 3.6.3)

4.3.9 Insertion Loss

The power transmission through a device is similarly defined in terms of the insertion loss, the ratio of output to input power expressed in dBs when all the ports except port j are terminated into reflectionless matched loads:

$$LA_{ij} = -10 \log_{10} \frac{P_{\text{out}}}{P_{\text{in}}} = -20 \log_{10} |\underline{s}_{ij}| \quad [\text{dB}]. \tag{4.22}$$

The input port is port j, the output port is port i. Depending on the particular choice of ports of a device, the insertion loss may also be called attenuation, coupling, or isolation (Chapter 6).

4.4 PROPERTIES OF THE DEVICES

4.4.1 Remark

A rigorous determination of the fields within a microstrip device may only be carried out for a limited number of simple geometries, even when using the most sophisticated computation techniques described in Chapter 10.

The scattering matrix of an n-port contains n^2 complex terms. Geometrical and physical properties provide relationships between some of them, allowing one to reduce the number of independent variables and, thus, the number of computations or measurements required to characterize the device.

4.4.2 Linearity

By writing a set of n equations with n variables, as is done in the definition of all the matrices presented in previous sections, one more or less assumes that the terms are independent from the amplitudes of the signals (i.e., that the device is linear). As a rule, matrix formulations are defined specifically for *linear devices*.

To some extent, matrices can also describe some nonlinear devices. For instance, in a frequency multiplier (a highly nonlinear device) the output signal at the angular frequency $k\omega$ may be linearly related to the input signal at the angular frequency ω. More generally, a nonsinusoidal signal may be represented by a combination of sine waves (Fourier series, intermodulation products, etc.), and for some well-behaved nonlinear devices these components are linearly related to each other from input to output. It is then possible to define an extended matrix formulation, with ports attributed to each frequency component. As an example, a 2-port in which six frequency components are considered becomes an equivalent 12-port device. Particular precautions must be taken when dealing with nonlinear devices.

4.4.3 Reciprocity

All devices made with dielectric and metal satisfy the reciprocity theorem, and in this case the transfer function does not depend upon the direction of propagation; thus,

$$\underline{S}_{ji} = \underline{S}_{ij}. \tag{4.23}$$

The number of independent terms in the scattering matrix is reduced to $n(n - 1)/2$. Reciprocity is not satisfied in devices containing magnetized ferrites (isolators, circulators, Section 6.7) and in most active devices (Chapter 8).

4.4.4 Symmetry

When a reciprocal device possesses one or several planes of geometric symmetry and when, in addition, the reference planes are symmetrically located, the corresponding terms of the matrices are either equal or of opposite sign, depending on the orientation of the fields. Symmetry is encountered in junctions and couplers (Chapter 6).

4.4.5 Reflectionless Match

A device is said to be *matched* at its port i when no signal leaves this port while all the other ports are terminated into reflectionless loads ($\underline{a}_j = 0$ for $j \neq i$). This corresponds to

$$\underline{s}_{ii} = 0 \quad \text{and} \quad \text{VSWR}_i = 1. \tag{4.24}$$

A device is matched at all its ports when all the terms on the diagonal of the scattering matrix are equal to zero.

4.4.6 Losslessness

In a lossless passive circuit, no energy is dissipated or produced, so the sum of all incoming power is equal to the sum of all outgoing power. This condition for energy conservation is given by

$$(\tilde{\underline{s}})(\underline{s}) = (1). \tag{4.25}$$

A device that does not dissipate any power cannot contain any resistive component. All the terms in the impedance and in the admittance matrices must be purely imaginary:

$$\underline{Z}_{ij} = jX_{ij} \quad \text{and} \quad \underline{Y}_{ij} = jB_{ij}. \tag{4.26}$$

4.5 EQUIVALENT CIRCUITS

4.5.1 Definition

Reciprocal devices can be represented by an equivalent circuit, which is a combination of inductors, capacitors, coupled coils, resistors, and other devices, whose frequency-dependent behavior simulates the behavior of the actual device. Equivalent circuits are mostly used for two-port devices.

4.5.2 Equivalent Π Network

The equivalent Π network for a two-port device is directly based on the terms of its admittance matrix, as shown in Figure 4.6.

4.5.3 Equivalent T Network

Conversely, the equivalent T network for a two-port device is directly based on the terms of its impedance matrix, also shown in Figure 4.6.

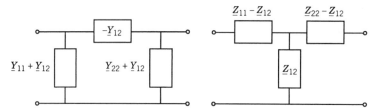

FIGURE 4.6   Π and T equivalent networks.

4.6   TRANSMISSION LINE SECTION

4.6.1   Equivalent Two-Port

The simplest possible "device" that one may consider is a section of lossless transmission line, of length L, defined by its characteristic impedance Z_c and its propagation factor β. The voltage and current along the line are given in Eq. 3.26 and are related to the device parameters at the two ends of the section by

$$\underline{U}_1 = \underline{U}(0), \qquad \underline{I}_1 = \underline{I}(0), \qquad \underline{U}_2 = \underline{U}(L), \qquad \underline{I}_2 = -\underline{I}(L). \quad (4.27)$$

4.6.2   Voltage – Current Related Matrices

The three matrices defined in terms of the voltages and currents at the ports are determined by means of simple algebraic derivations [Gardiol 1987]. The impedance matrix is

$$\begin{pmatrix} \underline{U}_1 \\ \underline{U}_2 \end{pmatrix} = \begin{pmatrix} \underline{Z}_{11} & \underline{Z}_{12} \\ \underline{Z}_{21} & \underline{Z}_{22} \end{pmatrix} \begin{pmatrix} \underline{I}_1 \\ \underline{I}_2 \end{pmatrix} = \begin{pmatrix} -jZ_c \cot \beta L & -jZ_c/\sin \beta L \\ -jZ_c/\sin \beta L & -jZ_c \cot \beta L \end{pmatrix} \begin{pmatrix} \underline{I}_1 \\ \underline{I}_2 \end{pmatrix}. \quad (4.28)$$

Similarly, one obtains the admittance matrix

$$\begin{pmatrix} \underline{I}_1 \\ \underline{I}_2 \end{pmatrix} = \begin{pmatrix} \underline{Y}_{11} & \underline{Y}_{12} \\ \underline{Y}_{21} & \underline{Y}_{22} \end{pmatrix} \begin{pmatrix} \underline{U}_1 \\ \underline{U}_2 \end{pmatrix} = \begin{pmatrix} (-j/Z_c)\cot \beta L & j/[Z_c \sin \beta L] \\ j/[Z_c \sin \beta L] & (-j/Z_c)\cot \beta L \end{pmatrix} \begin{pmatrix} \underline{U}_1 \\ \underline{U}_2 \end{pmatrix}, \quad (4.29)$$

while the chain or *ABCD* matrix is

$$\begin{pmatrix} \underline{U}_2 \\ -\underline{I}_2 \end{pmatrix} = \begin{pmatrix} A & B \\ C & D \end{pmatrix} \begin{pmatrix} \underline{U}_1 \\ \underline{I}_1 \end{pmatrix} = \begin{pmatrix} \cos \beta L & jZ_c \sin \beta L \\ (j/Z_c) \sin \beta L & \cos \beta L \end{pmatrix} \begin{pmatrix} \underline{U}_2 \\ \underline{I}_1 \end{pmatrix}. \quad (4.30)$$

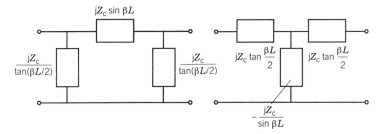

FIGURE 4.7   Π and T equivalent networks for a section of line.

It can readily be checked that the section of transmission line is linear, reciprocal, symmetrical, and lossless. Nothing may be said at this time about reflections, because these matrices do not contain any information about the lines to which the device may be connected.

4.6.3   Equivalent Circuits

The elements of the two equivalent circuits, in T and in Π, are determined respectively with the impedance and admittance matrices. The two circuits are presented in Figure 4.7.

These circuits represent a section of line of length L, while the equivalent circuit in Figure 3.4 is for a vanishingly short section of line of length $\mathrm{d}z$ (it may also be obtained by taking the limit $L \to \mathrm{d}z$).

4.6.4   Scattering Matrix

The scattering matrix of the section of line, inserted within connecting lines having the same characteristic impedance Z_c, is simply a diagonal phase-shift matrix:

$$\begin{pmatrix} \underline{b}_1 \\ \underline{b}_2 \end{pmatrix} = \begin{pmatrix} \underline{S}_{11} & \underline{S}_{12} \\ \underline{S}_{21} & \underline{S}_{22} \end{pmatrix} \begin{pmatrix} \underline{a}_1 \\ \underline{a}_2 \end{pmatrix} = \begin{pmatrix} 0 & \exp(-j\beta L) \\ \exp(-j\beta L) & 0 \end{pmatrix} \begin{pmatrix} \underline{a}_1 \\ \underline{a}_2 \end{pmatrix}. \quad (4.31)$$

Adding a section of lossless transmission line merely produces a phase shift. This was previously pointed out in Section 4.3.6.

4.6.5   Input Impedance

Let us consider the "usual" question, most often encountered when dealing with microstrip lines: what is the impedance at the input of a line connected to a load of impedance $\underline{Z}_L$ (Figure 4.8)?

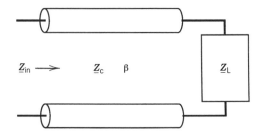

FIGURE 4.8 Section of line connected to a load.

Carrying out the calculations yields the input impedance

$$\underline{Z}_{in} = Z_c \frac{\underline{Z}_L + jZ_c \tan \beta L}{Z_c + j\underline{Z}_L \tan \beta L}.$$
(4.32)

Even though this is a quite common situation, the evaluation of the resulting complex expression is not straightforward. It becomes even more complex when the line is lossy [Gardiol 1987].

4.6.6 The Smith Chart

At a time when computers were yet to come, the computation of the input impedance with Eq. 4.29 led to severe practical problems during the development of early radars. Using graphical means, Philip Smith [1969] devised a

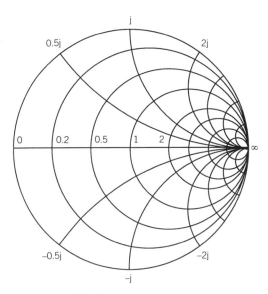

FIGURE 4.9 Smith chart.

most ingenious way to provide the desired value (Figure 4.9). Called the Smith chart, this graphical representation is still in use and appears routinely on the screens of microwave measuring instruments and computers.

The basic idea is to use the complex plane of the reflection factor ($\underline{s}_{ii}$) in which a translation along the line corresponds to a rotation around the center of the graph. A complete turn corresponds to a half-wavelength of line. The system of impedance coordinates, normalized with respect to the characteristic impedance of the line, is then projected onto the chart, providing a curvilinear set.

The use of the chart for the purpose of matching is described in next section.

4.7 MATCHING

4.7.1 Purpose

The presence of reflections on the connecting lines is generally troublesome:

The reflected signal does not contribute to the transmission to the load (receiver), its power is partly lost, resulting in an attenuation.

The generator may be affected by the signal returning from the load. As a result, the amplitude and the frequency of the signal produced may vary significantly ("frequency pulling").

The generator may further reflect part of the returned signal, setting up a regime of multiple reflections (echos). If the connecting line is long enough, the echos may be interpreted as new signals by the receiver, producing transmission errors.

The presence of reflected signals on a line produces local buildup of the fields, the amplitudes of which may become excessive in high-power applications.

To avoid these unwanted effects, one generally tries to reduce the reflected signal as much as possible by the process called *matching*.

4.7.2 Matching with Series Reactance

Considering Figure 4.8 and Eq. 4.32, notice that for some lengths L of transmission line the real part of the input impedance is equal to the characteristic impedance Z_c. The following expression gives the corresponding line lengths:

$$\tan \beta L = \frac{X_L Z_c \pm \sqrt{R_L Z_c \left[(R_L - Z_c)^2 + X_L^2 \right]}}{|\underline{Z}_L|^2 - R_L Z_c}, \qquad (4.33)$$

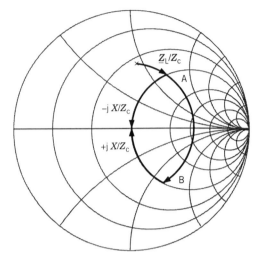

FIGURE 4.10   Reactive series match on the Smith chart.

where $\underline{Z}_L = R_L + jX_L$. It is then possible to completely suppress the reflection by adding a series reactance $-\text{Im}(\underline{Z}_{in})$ that exactly compensates the imaginary part of the input impedance.

The sequence of events is shown on the Smith chart in Fig. 4.10. A length of transmission line produces a clockwise rotation around the center of the chart until the corresponding point reaches the circle passing through the center, at points A and B. The addition of a reactance corresponds to a movement on this circle until the center of the chart is reached.

The length of line required is determined by the rotation angle, a full turn on the chart corresponding to a half-wavelength on the line (commercially available charts have a rotation scale). Matching from point A requires a negative reactance (i.e., a capacitance). From point B, a positive reactance (i.e., an inductance) is required to reach the center of the chart. Note that the point representing the load must not be on the outer circle of the Smith chart: the load impedance must have a resistive component, or else the load cannot absorb the signal.

4.7.3 Matching with Shunt Susceptance

In actual practice, the connection of series elements may be undesirable: one would have to cut across the line and then insert a reactive device (capacitor or inductor) across the gap. In microstrip, it is generally more convenient to realize shunt reactances, particularly capacitive stubs (Section 5.2.1).

The matching procedure can be used with admittances instead of impedances. An interesting property of the Smith chart is that the inverse of a complex number is the symmetrical point, just across the center of the

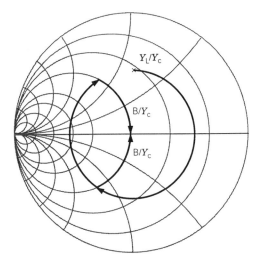

FIGURE 4.11   Reactive shunt match on the Smith chart.

chart. Therefore, the line lengths required are the values obtained in the previous section ± 1 quarter-wavelength. The matching process for the situation considered in Section 4.7.2 is sketched in Figure 4.11. The values on the chart correspond in this case to $\underline{Y}/Y_c$ (and not to $\underline{Z}/Z_c$ as in the previous section).

4.7.4   Matching with a Quarter-Wave Transformer

When a load impedance is purely real $(Z_L = R_L)$, it can be matched by means of a transmission line section $\lambda/4$ long having a characteristic impedance $Z_T = \sqrt{Z_c R_L}$ (Figure 4.12). Quarter-wave transformers are generally used to connect lines of different impedances. They can also be used to match complex loads by first introducing a section of transmission line, at the input of which the impedance is real.

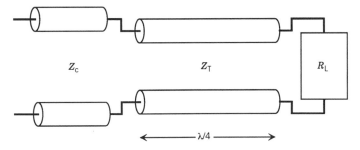

FIGURE 4.12   Matching with quarter-wave transformer.

4.7.5   Matching with Generalized Transformer

Some complex loads can be directly matched by a section of transmission line of unknown characteristic impedance Z_c and length L: to find these values, one sets Eq. 4.32 equal to the desired input impedance and solves for Z_c and L.

4.7.6   Comments

Several schemes are available to match a load, as long as the load itself is not purely reactive (total reflection). The designer must select the most convenient one. Adding lumped elements is often undesirable, and some matching devices may take too much space.

Matching is a frequency-dependent process: the sections of line are defined with respect to the wavelength, which is frequency dependent, and the reactances used for matching also vary with frequency. This means that, while a perfect match can be obtained at a single frequency, limitations are encountered when matching over a frequency band. For narrow-band devices, the matching process is carried out at the center frequency and the reflection factors are calculated at the edges of the band. More complex schemes are required for broad-band matching, in which case a computer-aided design software is used to optimize the match (Chapter 12).

4.7.7   Other Matching Conditions

The previous sections considered reflectionless matching, a process in which one wishes to eliminate the reflections. Two other matching possibilities, in which reflections are not suppressed, are considered in particular situations:

1. Maximization of the power absorbed by the load. This situation occurs at the conjugate match, when the input impedance is the complex conjugate of the source impedance.
2. Reduction of the noise figure. In transistors, the lowest noise figure is obtained with a source impedance that does not correspond to the reflectionless match. This is a basic requirement for the front stage of sensitive receivers (Section 8.4.4).

4.8   DEEMBEDDING

4.8.1   Definition

The properties of microstrip circuits defined in the previous sections are "intrinsic"; that is, they define the structure independently of how it is connected to the outer world. Experimental values, on the other hand, generally contain other parameters because measurements are made in coaxial line or waveguides across connectors and transitions (Sections 1.5.4 and 14.1). It is then necessary to extract the properties of the device from the measured values: this process is known as *deembedding*. The same situation is encountered in some theoretical simulations, where the structure analyzed includes the connectors (feeds).

4.8.2   Description of the Problem

A microstrip circuit is connected to microstrip lines terminated by coaxial connectors in planes $1', 2', \ldots, n'$ (Figure 4.13). The deembedding process extracts from the global impedance matrix $[\underline{Z}_{\text{tot}}]$ the intrinsic impedance matrix $[\underline{Z}]$ of the unknown structure. The reference planes of the device itself $(1, 2, \ldots, n)$ are placed in locations where the higher-order modes excited by the device and by the connector are of negligible amplitude.

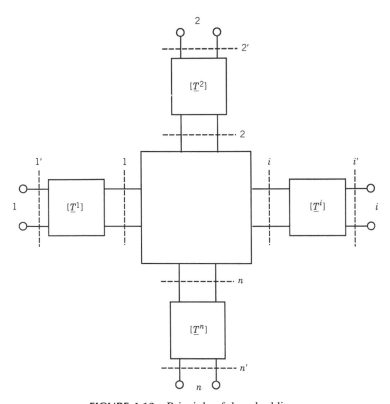

FIGURE 4.13   Principle of deembedding.

4.8.3 Mathematical Development

The global impedance matrix $[Z_{tot}]$, defined for the reference planes $1', 2', \ldots, n'$, is known either through measurements or simulation.

$$[u'] = [Z_{tot}][i'].\qquad(4.34)$$

The connecting lines, connectors, transitions are defined by two-port matrices $[Z^k]$, with $k = 1, 2, \ldots, n$, which must also be known. The corresponding chain matrices $[T^k]$ are specified at all the ports for the current directions shown in Figure 4.13 (Section 4.2.5) by

$$\left(\frac{U_k'}{I_k'}\right) = \left(\begin{matrix} T_{11}^k & T_{12}^k \\ T_{21}^k & T_{22}^k \end{matrix}\right)\left(\frac{U_k}{I_k}\right), \qquad k = 1, \ldots, n.\qquad(4.35)$$

The intrinsic impedance matrix of the microstrip circuit $[Z]$ is defined between the planes $1, 2, \ldots, n$ by

$$[u] = [Z][i].\qquad(4.36)$$

Four $n \times n$ diagonal matrices are defined:

$$[\mathrm{Diag}\, T_{ij}] = \begin{bmatrix} T_{ij}^1 & 0 & \cdots & 0 & \cdots & 0 \\ 0 & T_{ij}^2 & \cdots & 0 & \cdots & 0 \\ \cdots & \cdots & \cdots & \cdots & \cdots & \cdots \\ 0 & 0 & \cdots & T_{ij}^k & \cdots & 0 \\ \cdots & \cdots & \cdots & \cdots & \cdots & \cdots \\ 0 & 0 & \cdots & 0 & \cdots & T_{ij}^n \end{bmatrix}, \qquad i, j = 1, 2.\qquad(4.37)$$

After some developments, the intrinsic impedance matrix of the microstrip circuit is obtained:

$$[Z] = \{[\mathrm{Diag}\, T_{11}] - [Z_{tot}][\mathrm{Diag}\, T_{21}]\}^{-1}$$
$$\times \{[Z_{tot}][\mathrm{Diag}\, T_{22}] - [\mathrm{Diag}\, T_{12}]\}.\qquad(4.38)$$

4.9 PROBLEMS

4.9.1 A two-port device is represented by its impedance matrix, with $Z_{11} = Z_{22} = j35$ and $Z_{12} = Z_{21} = -j75$. Determine the corresponding admittance matrix, chain matrix, and the Π and T equivalent circuits.

4.9.2 A two-port device is represented by its scattering matrix, with $\underline{s}_{11} = \underline{s}_{22} = 0.6$ and $\underline{s}_{12} = \underline{s}_{21} = -j0.8$. Determine the corresponding impedance matrix and chain matrix when the device is connected to 50-Ω lines at both ports.

4.9.3 Indicate whether the devices represented by the following scattering matrices are reciprocal, matched, or lossless:

$$\begin{bmatrix} j0.5 & 0.707 & -j0.5 \\ 0.707 & 0 & 0.707 \\ -j0.5 & 0.707 & j0.5 \end{bmatrix} \begin{bmatrix} 0.1 & 0.99 & -0.1 \\ -0.1 & 0.1 & 0.99 \\ 0.99 & -0.1 & 0.1 \end{bmatrix} \begin{bmatrix} 0 & 0.5 & 0.5 \\ 0.5 & 0 & 0.5 \\ 0.5 & 0.5 & 0 \end{bmatrix}$$

4.9.4 Determine the impedance and chain matrices, the Π and T equivalent circuits for 50-Ω lossless transmission line segments of lengths $\lambda/8$, $\lambda/4$, and $\lambda/2$.

4.9.5 Show that when taking the limit $L \rightarrow dz$, the Π and T equivalent circuits of a lossy transmission line both yield the equivalent circuit of an elementary section of line.

4.9.6 We want to match a load of impedance $\underline{Z}_L = 146 + j13.8 \ \Omega$ to a transmission line with a characteristic impedance of 50 Ω. Determine the lengths of transmission lines and the reactive elements required for the four possible situations (series or shunt, inductive or capacitive).

4.9.7 A load has an impedance of $\underline{Z}_L = 18.4 + j28 \ \Omega$ and is to be matched with a transformer to a line of characteristic impedance $Z_c = 71.2 \ \Omega$. Determine the lengths of the line sections and the impedance for a quarter-wave transformer and for a generalized transformer.

CHAPTER FIVE

Discontinuities

A printed circuit is made of lines that are neither straight nor infinite. They begin somewhere and end elsewhere, their directions and their widths vary, they divide into multiple branches, they cross one another, and so on. Some discontinuities are shown in Figure 5.1. Discontinuities produce reflections of the signal and some radiation.

5.1 EQUIVALENT CIRCUITS

5.1.1 Higher-Order Modes

Next to a discontinuity, the boundary conditions (Eq. 2.6) are not satisfied by the field components of the dominant mode. Several modes of the line must be present to satisfy the continuity requirements. A discontinuity is thus a "source" of higher-order modes (Section 4.1.2).

5.1.2 Capacitance

Electric charges tend to pile up at the edges of a discontinuity, producing a local increase of charge on the conductor. The electric field increases and electric energy is stored. The resulting effect is capacitive, and one inserts a capacitor in the equivalent circuit to represent it. Similarly, dielectric losses within the substrate require the adjunction of a conductance.

5.1.3 Inductance

A change in the cross section of the conductor modifies the distribution of the current, producing a local increase of the magnetic field. Magnetic energy is then stored in the higher-order modes, producing an inductive effect

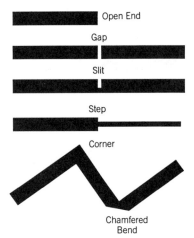

FIGURE 5.1 Discontinuities most often en- countered in microstrip.

represented by an inductance in the equivalent circuit. Additional conductor losses are similarly accounted for by a resistor.

5.1.4 LC Circuits

Discontinuities that modify at the same time the charge and the current densities are represented by a combination of equivalent LC components. The L and C components are frequency independent for low frequencies (static and quasistatic models), but become frequency dependent as frequency increases.

5.1.5 Continuous Conductor

When the upper conductor is continuous (Figure 5.2), the equivalent T network consists of two inductors and a capacitor. This is a low-pass filter, whose attenuation is vanishingly small at dc, but increases sharply for frequencies above its cutoff.

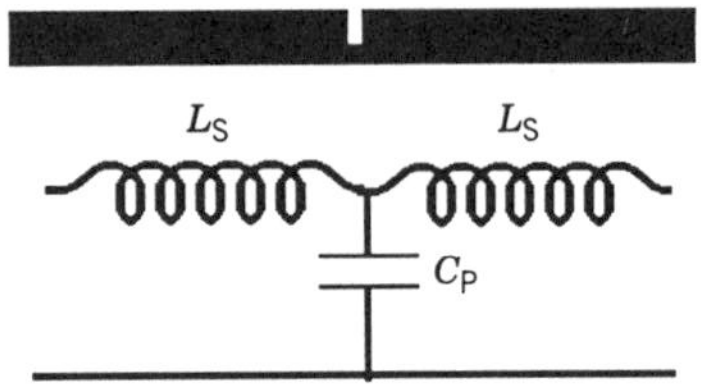

FIGURE 5.2 Discontinuity with continuous con- ductor.

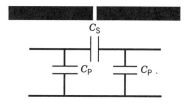

FIGURE 5.3 Discontinuity with discontinuous conductor.

5.1.6 Discontinuous Conductor

When there is no metallic connection between the input and output ports (Figure 5.3), direct current cannot cross the gap, which is then a dc block. The equivalent Π network is formed with capacitors only and is thus a capacitive voltage divider.

5.1.7 Remark

Reactive elements can also be represented by equivalent transmission line segments [Hoffmann 1987]. At high frequencies, the scattering matrix (Section 5.2) is preferred to characterize a discontinuity [Mehran 1976].

5.2 STUBS, GAPS, AND SLOTS

5.2.1 The Open Microstrip Line

Open microstrip lines are commonly used in practice for matching purposes (stubs). They were studied quite extensively by many authors, at first with low-frequency approximations [Farrar and Adams 1972; Silvester and Benedek 1973; James and Tse 1972; Jansen 1981] and later with full-wave solutions [Jackson and Pozar 1985; Jansen 1985; Katehi and Alexopoulos 1985; Koster and Jansen 1986; Jackson 1989].

At the open end of a transmission line, the fields do not stop abruptly but extend slightly further (fringing field). This effect can be modeled either with an equivalent capacitance or with an equivalent length of line Δl, the latter being more convenient in actual practice. An approximate expression was derived for Δl by Hammerstad and Bekkadal [1975], providing an accuracy better than 4% for $w/h \geq 0.2$ and $2 \leq \varepsilon_r \leq 50$:

$$\frac{\Delta l}{h} = 0.412 \frac{\varepsilon_e + 0.3}{\varepsilon_e - 0.258} \frac{w/h + 0.262}{w/h + 0.813}. \tag{5.1}$$

A more accurate approximation was established by Kirschning et al. [1981], improving the accuracy to better than 0.2% over the range $0.01 \leq$

$w/h \le 100$ and $\varepsilon_r \le 128$:

$$\frac{\Delta l}{h} = \frac{\xi_1 \xi_3 \xi_5}{\xi_4}, \tag{5.2}$$

where

$$\xi_1 = 0.434907 \frac{\varepsilon_e^{0.81} + 0.26}{\varepsilon_e^{0.81} - 0.189} \frac{(w/h)^{0.8544} + 0.236}{(w/h)^{0.8544} + 0.87},$$

$$\xi_2 = 1 + \frac{(w/h)^{0.371}}{2.358\varepsilon_r + 1},$$

$$\xi_3 = 1 + 0.5274 \frac{\tan^{-1}\{0.084(w/h)^{1.9413/\xi_2}\}}{\varepsilon_e^{0.9236}},$$

$$\xi_4 = 1 + 0.0377 \tan^{-1}\{0.067(w/h)^{1.456}\}[6 - 5\exp\{0.036(1 - \varepsilon_r)\}],$$

$$\xi_5 = 1 - 0.218\exp(-7.5w/h). \tag{5.3}$$

Radiation becomes significant when the substrate is thick and the strip is wide [James and Henderson 1979; Boukamp and Jansen 1984; Jackson and Pozar 1985; Katehi and Alexopoulos 1985].

5.2.2   The Gap in a Microstrip Line

The equivalent circuit of a microstrip gap is a Π network with a capacitance C_s in the series arm and two capacitances, C_{p1} and C_{p2}, in the shunt arms (Figure 5.2). Gaps were analyzed by a quasi-static method [Benedek and Silvester 1982; Maeda 1972] and a resonator method [Kirschning et al. 1983].

A full-wave analysis of microstrip gaps was carried out by Katehi and Alexopoulos [1985]. The effects of radiation and surface waves are taken into account by adding series and shunt conductances in parallel with the capacitances.

5.2.3   Transverse Slit in a Microstrip Line

The equivalent circuit of a notch or slot was determined with a simplified version of the waveguide model [Hoefer 1977]. When the reference planes are taken on its symmetry axis, the slit can be represented by a pure series inductance of value

$$\frac{L}{h} [\mu\text{H/m}] = 2\left(1 - \frac{Z_c}{Z_c'}\sqrt{\frac{\varepsilon_e}{\varepsilon_e'}}\right)^2, \tag{5.4}$$

where Z_c and ε_e are the characteristic impedance and effective permittivity of the microstrip line, while the primed quantities correspond to the reduced width section.

5.3 IMPEDANCE STEPS

Steps in width are routinely used in microstrip circuits to realize quarter-wave transformers and, thus, have received considerable attention from many authors [Benedek and Silvester 1972; Gupta et al. 1979; Garg and Bahl 1978; Koster and Jansen 1986; Gopinath et al. 1976]. Within the plane of the discontinuity, the step is represented by a T network with inductances in the series arms and a capacitance in the shunt arm (Figure 5.4).

For a symmetrical step the inductances are given by

$$L_1 = \frac{L_1'}{L_1' + L_2'} L, \qquad L_2 = \frac{L_2'}{L_1' + L_2'} L, \qquad (5.5)$$

where L_1' and L_2' are the inductances per unit length of the two lines, and

$$\frac{L}{h} \ [\text{nH/m}] = 40.5\left(\frac{w_2}{w_1} - 1\right) - 32.57 \ln\left(\frac{w_2}{w_1}\right) + 0.2\left(\frac{w_2}{w_1} - 1\right)^2, \quad (5.6)$$

with an accuracy of 5% or better for $w_2/w_1 \le 5$ and $w_2/h = 1$.

The step capacitance is given by

$$\frac{C}{\sqrt{w_1 w_2}} \ [\text{pF/m}] = (4.386 \ln \varepsilon_r + 2.33)\frac{w_2}{w_1} - 5.472 \ln \varepsilon_r - 3.17. \quad (5.7)$$

The error is smaller than 10% over the range $\varepsilon_r \le 10$ and $1.5 \le w/h \le 3.5$.

Tapered transitions were analyzed by Chadha and Gupta [1982]. The transmission line matrix (TLM) method was also applied to study discontinuities like impedance steps within completely closed structures [Akhtarzad and Johus 1975b].

5.4 MICROSTRIP BENDS

5.4.1 Right-Angle Bend or Corner

Bends are required in all circuits where the direction of the line changes, most often by an angle of 90°. In general, the strip widths are the same (equal impedances) on both sides of the bend, which presents therefore a plane of symmetry. Bends are represented by an equivalent T network (Figure 5.2), with inductances in the series arms and a capacitor in the shunt arm. The quasi-static bend capacitance C was determined by Silvester and

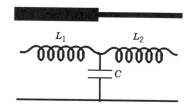

FIGURE 5.4   Microstrip step.

Benedek [1973], the inductance L by Thomson and Gopinath [1975]. Closed-form expressions for both quantities were established by Gupta et al. [1979].

$$\frac{C}{w} \ [\text{pF/m}] = \frac{(14\varepsilon_r + 12.5)w/h - (1.83\varepsilon_r - 2.25)}{\sqrt{w/h}} \qquad \text{for } \frac{w}{h} < 1,$$

$$\frac{C}{w} \ [\text{pF/m}] = (9.5\varepsilon_r + 1.25)\frac{w}{h} - 5.2\varepsilon_r + 7 \qquad \text{for } \frac{w}{h} > 1,$$

$$\tag{5.8}$$

$$L/h \ [\text{nH/m}] = 100\left(4\sqrt{w/h} - 4.21\right). \tag{5.9}$$

The accuracy on the capacitance is about 5% over the range $2.5 \leq \varepsilon_r \leq 15$ and $0.5 \leq w/h \leq 2$, while that on the inductance is about 3% over the range $0.5 \leq w/h \leq 2$ (compared with the results of Thomson and Gopinath 1975).

5.4.2   Chamfered Bend

The reflection produced by a corner can be reduced by rounding or chamfering it. With a Green's function in an integral equation formulation, Anders and Arndt [1980] determined that mitered bends tend to be more favorable than rounded ones. Dimensioning rules based on experimental data were established by Douville and James [1978]. The compensation over the range $1 \leq \varepsilon_r \leq 25$ and $w/h \geq 0.25$ is approximately optimal when

$$s/d = 0.52 + 0.65\exp(-1.35w/h), \tag{5.10}$$

where s and d are defined in Figure 5.5. The frequency-dependent scattering

FIGURE 5.5   Chamfered bend.

parameters of the mitered bend were determined with the waveguide model [Menzel 1976].

5.4.3   Radiation from Bends

Corners and bends tend to radiate and to excite surface waves when the fh product becomes large. Integral methods were used to analyze bends with different shapes [Skrivervik and Mosig 1990]. It was found that corners could radiate up to several percent of the incoming power, but that radiation is significantly reduced by chamfering, while rounded corners provide intermediate values. Radiation and reflection from bends were found to be practically correlated. This ranking was confirmed experimentally [Rautio and Harrington 1987].

An FDTD analysis of microstrip bends separated the effects of radiation and of surface waves, showing that considerable power can be lost by the two mechanisms [Feix et al. 1992].

5.5   REMARKS AND COMMENTS

One might expect that the results given by frequency-sensitive models, which are more complete in nature, would be more accurate. This is not always the case, and it was noticed that extrapolation of high-frequency results toward low frequencies sometimes led to erroneous values [Easter 1975].

For most low-frequency applications, quasi-static methods provide adequate information. Their main advantage lies in their simplicity, which results in a relatively small computational effort, and they allow easy synthesis of microstrip structures. Hoffmann [1987] summarized these techniques and applications with extensive references in his comprehensive book.

However, when microstrip structures are used at higher frequencies, the equivalent circuits based on a quasi-static approximation are no longer adequate. When transverse dimensions and substrate thickness become significant fractions of a wavelength, only full-wave techniques take into account the dispersion and the coupling produced by radiation and surface waves.

The descriptions of discontinuities in terms of their equivalent circuits often yield rather complex formulations, and only a few simple expressions are presented here. A large catalog of expressions is provided in Hoffmann's handbook [1987]. Computer techniques are constantly improved, providing more accurate approximations, and it is necessary to keep up-to-date with the latest publications. Many CAD packages (Chapter 12) upgrade the software on a more or less periodical basis to include the latest developments.

5.6 PROBLEMS

5.6.1 Calculate the equivalent increase in length for an open microstrip line in the case of 50-Ω lines on substrates 0.6 mm thick for relative permittivities of 1.5, 2, 3.5, 10, and 16.

5.6.2 Evaluate the components of the equivalent circuit of an impedance step from 50- to 80-Ω lines on a substrate with a thickness of 0.5 mm and a relative permittivity of 3.5.

5.6.3 Determine the chamfer required in a compensated mitered bend for a 0.9-mm-wide line on a 0.4-mm-thick substrate with a relative permittivity of 4.5.

CHAPTER SIX

Couplers and Junctions

Couplers and junctions, with three, four, and more ports, are used to distribute a signal between several channels or, conversely, to combine several signals. They can be used to pick up a small part of a signal for the purpose of monitoring its amplitude and frequency. They can also be used for matching purposes.

6.1 THREE-PORTS

6.1.1 Scattering Matrix and Signal Flow Graph

The scattering matrix of a three-port device contains nine terms:

$$
\begin{pmatrix} \underline{b}_1 \\ \underline{b}_2 \\ \underline{b}_3 \end{pmatrix} = \begin{pmatrix} \underline{s}_{11} & \underline{s}_{12} & \underline{s}_{13} \\ \underline{s}_{21} & \underline{s}_{22} & \underline{s}_{23} \\ \underline{s}_{31} & \underline{s}_{32} & \underline{s}_{33} \end{pmatrix} \begin{pmatrix} \underline{a}_1 \\ \underline{a}_2 \\ \underline{a}_3 \end{pmatrix}. \tag{6.1}
$$

The corresponding signal flow graph is shown in Figure 6.1.

6.1.2 The Inherent Mismatch of Lossless Reciprocal Three-Ports

Consider a power divider made by a reciprocal junction between lossless transmission lines: the reciprocity conditions (Section 4.4.3) are introduced

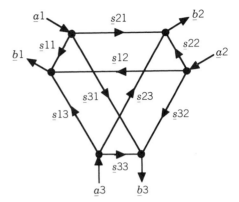

FIGURE 6.1 Flow graph of a three-port.

into Eq. 6.1, and the losslessness condition (Eq. 4.22) is evaluated. In addition, all the ports are assumed to be matched; that is, $\underline{s}_{11} = \underline{s}_{22} = \underline{s}_{33} = 0$. We then obtain the following relationships:

$$|\underline{s}_{12}|^2 + |\underline{s}_{13}|^2 = 1, \qquad \underline{s}_{12}^* \underline{s}_{13} = 0,$$

$$|\underline{s}_{12}|^2 + |\underline{s}_{23}|^2 = 1, \qquad \underline{s}_{12}^* \underline{s}_{23} = 0,$$

$$|\underline{s}_{13}|^2 + |\underline{s}_{23}|^2 = 1, \qquad \underline{s}_{13}^* \underline{s}_{23} = 0. \tag{6.2}$$

Let us assume that one of the terms, say $\underline{s}_{13}$, is nonzero. Then, considering the right-hand expressions, we find that one must have $\underline{s}_{12} = \underline{s}_{23} = 0$, which is clearly inconsistent with the second line in the left-hand expressions. This means that it is not possible to match at the same time the three ports of a lossless symmetrical three-port. For a three-port to be matched, it must be either lossy (Sections 6.2.3 and 6.2.4) or nonreciprocal (Section 6.3).

In a similar fashion, one may prove that, when matching two of the three ports, the third port must be completely decoupled from the other two (Problem 6.8.1).

6.1.3 Remark

In actual practice, matching does not really mean setting the diagonal terms of the matrix exactly equal to zero, but reducing their amplitude to an "acceptably low" value.

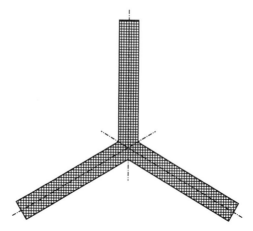

FIGURE 6.2 Symmetrical Y-junction.

6.1.4 Symmetrical Y-Junction

The threefold symmetry of the Y-junction (Figure 6.2) and the reciprocity condition reduce the number of terms of the scattering matrix to only two:

$$\underline{S}_{11} = \underline{S}_{22} = \underline{S}_{33} = A,$$

$$\underline{S}_{12} = \underline{S}_{13} = \underline{S}_{23} = \underline{S}_{21} = \underline{S}_{31} = \underline{S}_{32} = B + jC. \qquad (6.3)$$

The reference planes are selected so that the diagonal terms are real (to simplify further developments). The lossless condition then yields two equations:

$$A^2 + 2B^2 + 2C^2 = 1,$$

$$2AB + B^2 + C^2 = 0. \qquad (6.4)$$

The two limiting situations are

1. $A = 1$, where reflections are complete at all ports and no transmission takes place across the junction.
2. $A = \frac{1}{3}$, which corresponds to a VSWR of 2. The transmission factor to the two output ports is then $-\frac{2}{3} + j0$, providing an insertion loss of 3.52 dB. This situation corresponds to a connection of one line to two lines having the same characteristic impedance without reactive elements.

Y-junctions were analyzed by means of the equivalent waveguide method, taking into account the presence of higher-order modes [Menzel 1978].

6.2 POWER DIVIDERS

6.2.1 Lossless Power Divider

In a power divider, one wishes to have a matched input and a specified distribution of the input signal between the two outputs. With lines terminated into adequately matched loads, no signals are reflected, so it does not matter that the output ports cannot be matched.

A divider made by the parallel connection of three lines is represented by the equivalent circuit of Figure 6.3, in which higher-order modes are taken into account by a parallel susceptance jB.

The scattering matrix is then provided by Eq. 4.16:

$$(\underline{s}) = \frac{1}{Y_1 + Y_2 + Y_3 + jB}$$

$$\times \begin{pmatrix} Y_1 - Y_2 - Y_3 - jB & 2\sqrt{Y_1 Y_2} & 2\sqrt{Y_1 Y_3} \\ 2\sqrt{Y_1 Y_2} & Y_2 - Y_1 - Y_3 - jB & 2\sqrt{Y_2 Y_3} \\ 2\sqrt{Y_1 Y_3} & 2\sqrt{Y_2 Y_3} & Y_3 - Y_1 - Y_2 - jB \end{pmatrix}.$$

$$(6.5)$$

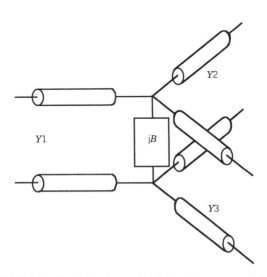

FIGURE 6.3   Equivalent circuit of a lossless power divider.

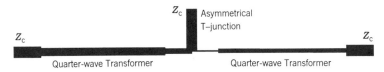

FIGURE 6.4   Power divider with output transformers.

The matching condition then requires that

$$Y_1 = Y_2 + Y_3 \quad \text{and} \quad B = 0, \tag{6.6}$$

and the scattering matrix takes the form

$$(\underline{s}) = \begin{pmatrix} 0 & \sqrt{\dfrac{Y_2}{Y_1}} & \sqrt{\dfrac{Y_3}{Y_1}} \\[2ex] \sqrt{\dfrac{Y_2}{Y_1}} & -\dfrac{Y_3}{Y_1} & \dfrac{\sqrt{Y_2 Y_3}}{Y_1} \\[2ex] \sqrt{\dfrac{Y_3}{Y_1}} & \dfrac{\sqrt{Y_2 Y_3}}{Y_1} & -\dfrac{Y_2}{Y_1} \end{pmatrix}. \tag{6.7}$$

The power division is directly related to the admittance ratio:

$$\frac{P_2}{P_3} = \frac{s_{21}^2}{s_{31}^2} = \frac{Y_2}{Y_3}. \tag{6.8}$$

A symmetrical divider provides an equal power split between the two outputs, with an insertion loss of 3 dB. Since the output ports connect to lines having different characteristic admittances, the device is completed by quarter-wave transformers (Figure 6.4). In this fashion, one can then connect the outputs to lines having the standard impedance (generally 50 Ω).

The device is reciprocal so that the direction of propagation of the signals can be reversed and the same behavior will be observed. This means that if two signals in phase and with adequate power ratio enter ports 2 and 3 of the power divider, these signals will be entirely transmitted through port 1. Signals that do not meet the amplitude and phase requirements (unbalanced components) are reflected.

6.2.2   T-Junction

Power dividers generally make use of a T-junction, to which many publications have been devoted [Hoffmann 1987]. Analyses were made with integral

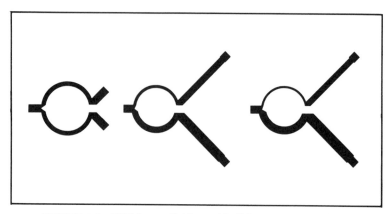

FIGURE 6.5 Wilkinson dividers with different power ratios.

techniques [Farrar and Adams 1972; Silvester and Benedek 1973; Wu et al. 1990] and with the waveguide model [Leighton and Milnes 1971; Wolff et al. 1972; Mehran 1975]. The cutting of a notch [Kompa 1976b] and other compensation techniques were reported [Dydyk 1977].

6.2.3 Wilkinson Divider

Refer to Figure 6.5. The output mismatches can be suppressed by connecting a resistor across the two output lines to absorb the unbalanced signal component (Section 6.2.1). The resulting divider is then no longer lossless.

Due to the presence of quarter-wave sections, the single-section Wilkinson divider is a relatively narrow-band device (Figure 6.6). Wider bandwidth operation can be obtained with multisection devices [Li et al. 1984].

6.2.4 Resistive Power Divider

In some applications, in particular within computer systems, any one of the three ports may become an input that should be matched for proper operation. A fully matched symmetrical divider requires resistors at its three ports (Figure 6.7).

The corresponding scattering matrix is given by

$$(\underline{s}) = \frac{1}{2} \begin{pmatrix} 0 & 1 & 1 \\ 1 & 0 & 1 \\ 1 & 1 & 0 \end{pmatrix}. \tag{6.9}$$

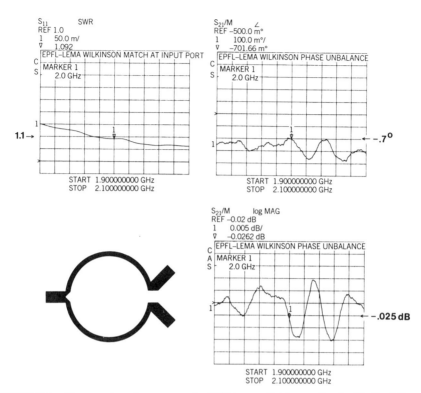

FIGURE 6.6 Measured performance of a Wilkinson divider realized with the Micros CAD package [Zürcher 1985].

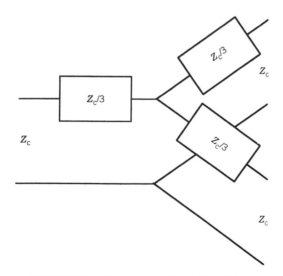

FIGURE 6.7 Fully matched resistive divider.

The insertion loss from any port to any other port is then 6 dB [Johnson 1975]. Note that one half of the signal power is dissipated within the divider.

6.3   CIRCULATOR

6.3.1   Matched Lossless Nonreciprocal Three-Port

The lossless conditions for a matched nonreciprocal three-port take the form

$$|\underline{s}_{21}|^2 + |\underline{s}_{31}|^2 = 1, \qquad \underline{s}_{12}^*\underline{s}_{13} = 0,$$

$$|\underline{s}_{12}|^2 + |\underline{s}_{32}|^2 = 1, \qquad \underline{s}_{21}^*\underline{s}_{23} = 0,$$

$$|\underline{s}_{13}|^2 + |\underline{s}_{23}|^2 = 1, \qquad \underline{s}_{31}^*\underline{s}_{32} = 0. \qquad (6.10)$$

There are now six equations with six unknowns, and the system admits the solution

$$(\underline{s}) = \begin{pmatrix} 0 & 0 & 1 \\ 1 & 0 & 0 \\ 0 & 1 & 0 \end{pmatrix}. \qquad (6.11)$$

The corresponding device is a circulator, whose flow graph and symbol are shown in figure 6.8. The signal entering port 1 leaves the device at port 2, the one entering port 2 leaves at port 3, and that entering port 3 leaves at port 1. The device is nonreciprocal, because an inversion of the sense of rotation produces a different device.

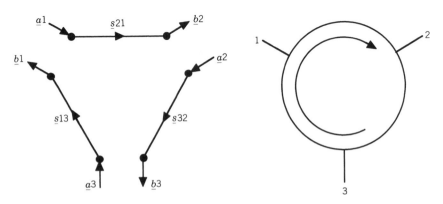

FIGURE 6.8   Flow graph and symbol of circulator.

6.3.2   Ferrite Circulator

Ferrites are ceramics that possess magnetic properties. They consist of a mixture of magnetic oxides of iron, nickel, cobalt, and rare-earth elements, with the addition of nonmagnetic metal oxides. In the presence of a biasing magnetic field, a ferrite material becomes magnetically anisotropic, that is, its permeability μ becomes a tensor: the material is called gyromagnetic.

When a symmetrical Y-junction is deposited on a ferrite substrate and a biasing magnetic field is applied perpendicularly to the plane of the interface, the gyromagnetic effect produces a rotation of the plane of polarization of the fields. In the absence of magnetic bias, the current spreads out symmetrically across the junction and couples equally to the two output ports. Applying a magnetic bias produces a rotation of the current distribution, and one of the ports can thus be decoupled. The sense of rotation remains the same when other ports are excited, so the device is a circulator [Fay and Comstock 1965; Bosma 1962].

The actual analysis of the fields within a ferrite substrate is extremely complex [Yang et al. 1992] and so is the design of microwave ferrite devices.

6.3.3   Applications of Circulators

Ferrite circulators are interconnecting components that separate different paths within a system. In radars, the circulator ports are connected to the transmitter, to the antenna, and to the receiver. The transmitter feeds the antenna, and the returned signal is fed to the receiver. Similarly, connecting a circulator to a reflection amplifier permits the separation of the input from the output.

Actual circulators are not ideal devices, in which a small part of the input signal is absorbed and another small part is transmitted to the "uncoupled" port. Typically, a good circulator provides an insertion loss (LA_{21}, Section 4.3.9) of 0.3–0.5 dB, with an isolation (LA_{12}) of 15–20 dB over the specified frequency band (the narrower the band, the better the performance).

The circulator function can also be realized with active circuits by combining amplifiers, sections of transmission lines, and reactive elements (Section 8.4.6).

6.3.4   Ferrite Isolator

Terminating one of the circulator's ports into a matched load produces an isolator, a device that transmits a signal in one direction but blocks it in the other. It is used to protect the signal generator from unwanted reflections and to avoid feedback between successive sections of a system.

6.4   FOUR-PORTS

6.4.1   Scattering Matrix and Signal Flow Graph

The scattering matrix of a four-port device contains 16 terms:

$$
\begin{pmatrix} \underline{b}_1 \\ \underline{b}_2 \\ \underline{b}_3 \\ \underline{b}_4 \end{pmatrix} = \begin{pmatrix} \underline{s}_{11} & \underline{s}_{12} & \underline{s}_{13} & \underline{s}_{14} \\ \underline{s}_{21} & \underline{s}_{22} & \underline{s}_{23} & \underline{s}_{24} \\ \underline{s}_{31} & \underline{s}_{32} & \underline{s}_{33} & \underline{s}_{34} \\ \underline{s}_{41} & \underline{s}_{42} & \underline{s}_{43} & \underline{s}_{44} \end{pmatrix} \begin{pmatrix} \underline{a}_1 \\ \underline{a}_2 \\ \underline{a}_3 \\ \underline{a}_4 \end{pmatrix} .
\tag{6.12}
$$

The corresponding signal flow graph is shown in Figure 6.9.

6.4.2   Matched Lossless Reciprocal Four-Port

When a four-port device is reciprocal, lossless, and matched, it can be shown that one output port is always isolated from the input port. The device obtained in this manner is called a *directional coupler*. When port 3 is decoupled from port 1, the scattering matrix becomes

$$
\begin{pmatrix} \underline{b}_1 \\ \underline{b}_2 \\ \underline{b}_3 \\ \underline{b}_4 \end{pmatrix} = \begin{pmatrix} 0 & \underline{s}_{12} & 0 & \underline{s}_{14} \\ \underline{s}_{12} & 0 & \underline{s}_{23} & 0 \\ 0 & \underline{s}_{23} & 0 & \underline{s}_{34} \\ \underline{s}_{14} & 0 & \underline{s}_{34} & 0 \end{pmatrix} \begin{pmatrix} \underline{a}_1 \\ \underline{a}_2 \\ \underline{a}_3 \\ \underline{a}_4 \end{pmatrix} .
\tag{6.13}
$$

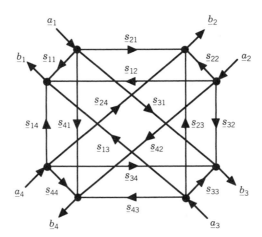

FIGURE 6.9   Flow graph of a four-port.

The matrix components must satisfy the following conditions:

$$|\underline{s}_{12}| = |\underline{s}_{34}| = \alpha \quad \text{and} \quad |\underline{s}_{14}| = |\underline{s}_{23}| = \beta \quad \text{with } \alpha^2 + \beta^2 = 1,$$

$$\underline{s}_{12}\underline{s}_{23}^* + \underline{s}_{14}\underline{s}_{34}^* = 0 \quad \text{and} \quad \underline{s}_{12}\underline{s}_{14}^* + \underline{s}_{23}\underline{s}_{34}^* = 0. \tag{6.14}$$

The reference planes are selected so that $\underline{s}_{12}$ and $\underline{s}_{34}$ are real, and the scattering matrix becomes

$$(\underline{s}) = \begin{pmatrix} 0 & \alpha & 0 & \beta \exp(j\psi) \\ \alpha & 0 & \beta \exp(j\theta) & 0 \\ 0 & \beta \exp(j\theta) & 0 & \alpha \\ \beta \exp(j\psi) & 0 & \alpha & 0 \end{pmatrix}, \tag{6.15}$$

with $\psi + \theta = \pi/2$. Two situations present particular symmetry conditions.

6.4.3 Symmetrical Coupler

When $\psi = \theta = \pi/4$, all the β terms are pure imaginary and the scattering matrix takes the form

$$(\underline{s}) = \begin{pmatrix} 0 & \alpha & 0 & j\beta \\ \alpha & 0 & j\beta & 0 \\ 0 & j\beta & 0 & \alpha \\ j\beta & 0 & \alpha & 0 \end{pmatrix}. \tag{6.16}$$

This situation can be obtained when the four-port exhibits a geometric symmetry and the reference planes are located symmetrically (branch line coupler, Section 6.5.1). It can also be obtained by selecting particular reference planes when the coupler has no particular geometric symmetry.

6.4.4 Antisymmetrical Coupler

For $\psi = 0$ and $\theta = \pi/2$, the terms in β are purely real, with positive and negative values, and the scattering matrix takes the form

$$(\underline{s}) = \begin{pmatrix} 0 & \alpha & 0 & \beta \\ \alpha & 0 & -\beta & 0 \\ 0 & -\beta & 0 & \alpha \\ \beta & 0 & \alpha & 0 \end{pmatrix}. \tag{6.17}$$

In this case, too, the antisymmetry may correspond to a geometric feature (Section 6.5.3) or to a particular choice of reference planes. Any symmetrical

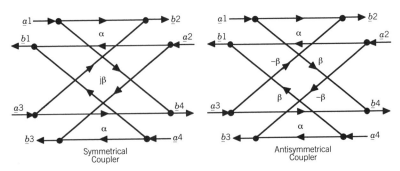

FIGURE 6.10 Flow graphs of symmetrical and antisymmetrical couplers.

coupler may be changed into an antisymmetrical coupler by modifying the location of the reference planes.

The flow graphs of symmetrical and antisymmetrical couplers are shown in Figure 6.10.

6.4.5 Symmetrical Four-Port

In this section, we consider a reciprocal four-port having two geometric planes of symmetry, as shown in Figure 6.11. At this point, the device is not assumed to be matched. The reference planes in the four ports are located symmetrically. From reciprocity and symmetry considerations, some components of the scattering matrix are equal to other ones and the device can be described by only four quantities:

$$\underline{S}_{11} = \underline{S}_{22} = \underline{S}_{33} = \underline{S}_{44} = \underline{S}_1, \qquad \underline{S}_{12} = \underline{S}_{21} = \underline{S}_{34} = \underline{S}_{43} = \underline{S}_2,$$

$$\underline{S}_{13} = \underline{S}_{31} = \underline{S}_{24} = \underline{S}_{42} = \underline{S}_3, \qquad \underline{S}_{14} = \underline{S}_{41} = \underline{S}_{23} = \underline{S}_{32} = \underline{S}_4. \qquad (6.18)$$

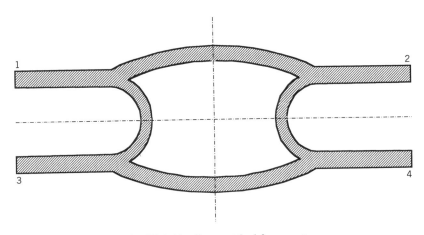

FIGURE 6.11 Symmetrical four-port.

The scattering matrix then takes the form

$$
\begin{pmatrix} \underline{b}_1 \\ \underline{b}_2 \\ \underline{b}_3 \\ \underline{b}_4 \end{pmatrix} = \begin{pmatrix} \underline{s}_1 & \underline{s}_2 & \underline{s}_3 & \underline{s}_4 \\ \underline{s}_2 & \underline{s}_1 & \underline{s}_4 & \underline{s}_3 \\ \underline{s}_3 & \underline{s}_4 & \underline{s}_1 & \underline{s}_2 \\ \underline{s}_4 & \underline{s}_3 & \underline{s}_2 & \underline{s}_1 \end{pmatrix} \begin{pmatrix} \underline{a}_1 \\ \underline{a}_2 \\ \underline{a}_3 \\ \underline{a}_4 \end{pmatrix}. \tag{6.19}
$$

6.4.6 Symmetrical and Antisymmetrical Excitations

Applying a doubly symmetrical excitation—signals of equal amplitude and phase at each of the four ports—one obtains

$$\underline{a}_1 = \underline{a}_2 = \underline{a}_3 = \underline{a}_4 = \underline{a}_{ss}$$
$$\underline{b}_1 = \underline{b}_2 = \underline{b}_3 = \underline{b}_4 = \underline{b}_{ss} = (\underline{s}_1 + \underline{s}_2 + \underline{s}_3 + \underline{s}_4)\underline{a}_{ss} = \underline{\rho}_{ss}\underline{a}_{ss}. \tag{6.20}$$

For a doubly antisymmetrical excitation, one similarly obtains

$$\underline{a}_1 = -\underline{a}_2 = -\underline{a}_3 = \underline{a}_4 = \underline{a}_{aa},$$
$$\underline{b}_1 = -\underline{b}_2 = -\underline{b}_3 = \underline{b}_4 = \underline{b}_{aa} = (\underline{s}_1 - \underline{s}_2 - \underline{s}_3 + \underline{s}_4)\underline{a}_{aa} = \underline{\rho}_{aa}\underline{a}_{aa}. \tag{6.21}$$

When the excitation is antisymmetrical left and right, but symmetrical top and bottom, the expressions become

$$\underline{a}_1 = -\underline{a}_2 = \underline{a}_3 = -\underline{a}_4 = \underline{a}_{as},$$
$$\underline{b}_1 = -\underline{b}_2 = \underline{b}_3 = -\underline{b}_4 = \underline{b}_{as} = (\underline{s}_1 - \underline{s}_2 + \underline{s}_3 - \underline{s}_4)\underline{a}_{as} = \underline{\rho}_{as}\underline{a}_{as}. \tag{6.22}$$

Finally, for an excitation symmetrical left and right and asymmetrical top and bottom, one gets

$$\underline{a}_1 = \underline{a}_2 = -\underline{a}_3 = -\underline{a}_4 = \underline{a}_{sa},$$
$$\underline{b}_1 = \underline{b}_2 = -\underline{b}_3 = -\underline{b}_4 = \underline{b}_{sa} = (\underline{s}_1 + \underline{s}_2 - \underline{s}_3 - \underline{s}_4)\underline{a}_{sa} = \underline{\rho}_{sa}\underline{a}_{sa}. \tag{6.23}$$

The four reflection factors obtained are related to the terms of the scattering matrix by

$$
\begin{pmatrix} \underline{\rho}_{ss} \\ \underline{\rho}_{as} \\ \underline{\rho}_{sa} \\ \underline{\rho}_{aa} \end{pmatrix} = \begin{pmatrix} 1 & 1 & 1 & 1 \\ 1 & -1 & 1 & -1 \\ 1 & 1 & -1 & -1 \\ 1 & -1 & -1 & 1 \end{pmatrix} \begin{pmatrix} \underline{s}_1 \\ \underline{s}_2 \\ \underline{s}_3 \\ \underline{s}_4 \end{pmatrix}. \tag{6.24}
$$

Conversely, the scattering matrix parameters can be determined with the four reflection factors:

$$\begin{pmatrix} \underline{S}_1 \\ \underline{S}_2 \\ \underline{S}_3 \\ \underline{S}_4 \end{pmatrix} = \frac{1}{4} \begin{pmatrix} 1 & 1 & 1 & 1 \\ 1 & -1 & 1 & -1 \\ 1 & 1 & -1 & -1 \\ 1 & -1 & -1 & 1 \end{pmatrix} \begin{pmatrix} \underline{\rho}_{ss} \\ \underline{\rho}_{as} \\ \underline{\rho}_{sa} \\ \underline{\rho}_{aa} \end{pmatrix}. \qquad (6.25)$$

6.4.7 Physical Meaning of the Anti- and Symmetrical Excitations

When two ports are excited symmetrically, the same currents are injected at both ports. On the plane of symmetry between the two ports, the two currents, directed along opposite directions, cancel each other so that no current crosses this plane: it is an open-circuit plane.

Conversely, when two ports are excited in antisymmetry, the voltages are opposed and cancel each other within the plane of symmetry, which is then a short-circuit plane. The four situations are sketched in Figure 6.12.

The complete analysis of a symmetrical four-port can thus be reduced to the analysis of one quarter of its structure, with the use of open-circuit and short-circuit terminating planes. Furthermore, when the device is lossless, the four reflection factors take unit amplitudes, so only their phases have to be determined.

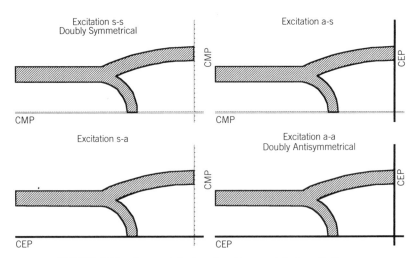

FIGURE 6.12   Symmetrical and antisymmetrical excitations.

6.4.8 Matched and Directive Four-Port

For the four-port to be adapted, one must have $\underline{s}_1 = 0$, and thus

$$\underline{\rho}_{ss} + \underline{\rho}_{as} + \underline{\rho}_{sa} + \underline{\rho}_{aa} = 0. \qquad (6.26)$$

As pointed out in Section 6.4.2, one port has to be decoupled from the input. Considering the same situation as previously, $\underline{s}_3 = 0$, and

$$\underline{\rho}_{ss} + \underline{\rho}_{as} - \underline{\rho}_{sa} - \underline{\rho}_{aa} = 0. \qquad (6.27)$$

These conditions are satisfied when

$$\underline{\rho}_{ss} = -\underline{\rho}_{as} \quad \text{and} \quad \underline{\rho}_{sa} = -\underline{\rho}_{aa}. \qquad (6.28)$$

The remaining terms of the scattering matrix are then

$$\underline{s}_2 = \tfrac{1}{2}(\underline{\rho}_{ss} - \underline{\rho}_{aa}) \quad \text{and} \quad \underline{s}_4 = \tfrac{1}{2}(\underline{\rho}_{ss} + \underline{\rho}_{aa}). \qquad (6.29)$$

The device is a symmetrical coupler. Selecting the reference planes as in Section 6.4.3, the terms α and β are then

$$\alpha = s_2 = \tfrac{1}{2}(\underline{\rho}_{ss} - \underline{\rho}_{aa}) \quad \text{and} \quad j\beta = \underline{s}_4 = \tfrac{1}{2}(\underline{\rho}_{ss} + \underline{\rho}_{aa}), \qquad (6.30)$$

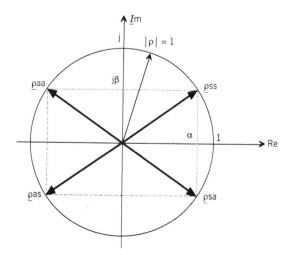

FIGURE 6.13 Location of the reflection factors.

and the four reflection factors become

$$\underline{\rho}_{ss} = -\underline{\rho}_{as} = \alpha + j\beta \quad \text{and} \quad \underline{\rho}_{aa} = -\underline{\rho}_{sa} = -\alpha + j\beta. \quad (6.31)$$

The four reflection factors are located at the four corners of a rectangle inscribed in the unit circle in the complex plane (Figure 6.13).

6.5 SHORT COUPLERS AND JUNCTIONS

6.5.1 Branch Line Coupler

A branch line coupler consists of two main transmission lines shunt-connected by two or more secondary lines or branch lines (Figure 6.14). This structure is a symmetrical four-port and, when adapted, a symmetrical coupler. The length and impedances of the lines, for a specified division ratio, can be determined by the technique outlined in the previous sections. The reflection factors for the four conditions of excitation are (neglecting reactive effects at the T-junction)

$$\underline{\rho}_{ss} = \frac{Y_c - jY_1 \tan \beta_1 l_1 - jY_2 \tan \beta_2 l_2}{Y_c + jY_1 \tan \beta_1 l_1 + jY_2 \tan \beta_2 l_2} \exp(j\phi) = \alpha + j\beta,$$

$$\underline{\rho}_{as} = \frac{Y_c + jY_1 \cot \beta_1 l_1 - jY_2 \tan \beta_2 l_2}{Y_c - jY_1 \cot \beta_1 l_1 + jY_2 \tan \beta_2 l_2} \exp(j\phi) = -\alpha - j\beta,$$

$$\underline{\rho}_{sa} = \frac{Y_c - jY_1 \tan \beta_1 l_1 + jY_2 \cot \beta_2 l_2}{Y_c + jY_1 \tan \beta_1 l_1 - jY_2 \cot \beta_2 l_2} \exp(j\phi) = \alpha - j\beta,$$

$$\underline{\rho}_{aa} = \frac{Y_c + jY_1 \cot \beta_1 l_1 + jY_2 \cot \beta_2 l_2}{Y - jY_1 \cot \beta_1 l_1 - jY_2 \cot \beta_2 l_2} \exp(j\phi) = -\alpha + j\beta. \quad (6.32)$$

FIGURE 6.14 Branch line coupler.

The phase ϕ was introduced to take into account the distance from the T-junction to the reference plane defined in the previous section.

The conditions of Eq. 6.28 require that

$$Y_1^2 = Y_c^2 + Y_2^2 \quad \text{and} \quad Y_1 \cot 2\beta_1 l_1 + Y_2 \cot 2\beta_2 l_2 = 0. \quad (6.33)$$

Applying Eq. 6.31, one further finds that $\phi = \pi/2$.

This problem admits a simple solution. Setting the two terms in the right part of Eq. 6.33 equal to zero gives

$$\beta l_1 = \beta l_2 = \pi/4, \qquad Y_1 = Y_c/\alpha, \qquad Y_2 = -Y_c\beta/\alpha. \quad (6.34)$$

In the present case, β is negative. The half-lengths of both lines are $\lambda_i/8$. In microstrip, the wavelength depends on the characteristic impedance, so the branch line coupler is not square. In the hybrid (equal split) coupler, $\alpha = \beta = 1/\sqrt{2}$; hence,

$$Y_1 = \sqrt{2}\,Y_c, \qquad Y_2 = Y_c. \quad (6.35)$$

The signal flow on a hybrid branch line coupler, showing the equal power split between the two ports on the right side and the canceling of the signals on the fourth port, is shown in Figure 6.15. This is actually a measured plot of the electric field normal to the air–dielectric interface.

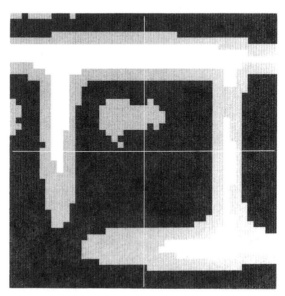

FIGURE 6.15   Distribution of the signal on a branch-line coupler. (Courtesy J-F. Zürcher.)

FIGURE 6.16   Different branch line coupler geometries.

Equations 6.32 admit other solutions, with line lengths different from $\lambda_i/8$ (one line shorter, the other longer). The characteristic admittances are then

$$Y_1 = \frac{Y_c}{\sqrt{1 - (\cot 2\beta_1 l_1/\cot 2\beta_2 l_2)^2}},$$

$$Y_2 = \frac{Y_c}{\sqrt{1 + (\cot 2\beta_2 l_2/\cot 2\beta_1 l_1)^2}}. \quad (6.36)$$

The scattering parameters are

$$\alpha = \frac{2Y_c(Y_1 \tan \beta_1 l_1 + Y_2 \tan \beta_2 l_2)}{Y_c^2 + (Y_1 \tan \beta_1 l_1 + Y_2 \tan \beta_2 l_2)^2},$$

$$\beta = \frac{Y_c^2 - (Y_1 \tan \beta_1 l_1 + Y_2 \tan \beta_2 l_2)^2}{Y_c^2 + (Y_1 \tan \beta_1 l_1 + Y_2 \tan \beta_2 l_2)^2} \quad (6.37).$$

In these developments, the reactive elements representing the effect of higher modes in the T-junction were neglected. To some extent, they can be reduced or compensated for by adjusting the geometry [Anada et al. 1991]. The directivity can be increased by using Y-junctions, but then the connecting lines are no longer in-line [Chadha 1981]. Several forms of branch line couplers are shown in Figure 6.16 [Wright and Judah 1987].

6.5.2   Broad-Band Branch Line Coupler

The line segments are $\lambda_i/8$ long at a single frequency only, and the performance of the device is degraded for signals at other frequencies. Broadbanding can be done by using multiple-section couplers [Matthaei et al. 1964], and other schemes have been proposed [Paul et al. 1991].

6.5.3   Rat Race Hybrid Ring

The rat race hybrid (Figure 6.17) is made of an annular microstrip line having a mean circumference of $1.5\lambda_g$ and a characteristic impedance of $Z_c\sqrt{2}$ and is connected to four transmission lines of characteristic impedance Z_c. Numbering the ports as indicated on the figure and taking reference planes at an equal distance from the junctions in all the ports, one obtains the scattering matrix of an antisymmetrical hybrid junction (Eq. 6.17 with $\alpha = \beta = 1/\sqrt{2}$).

A signal entering port 1 is distributed equally and in phase between ports 2 and 4, with no coupling to port 3. A signal fed to port 3 is divided equally, but in phase opposition, to ports 2 and 4, with no signal coming out of port 1. The signal flows for these two situations are shown in Figure 6.18 (measured plot of the electric field normal to the air–dielectric interface).

When two signals are fed to ports 2 and 4, the sum of the signals leaves port 1 and their difference leaves port 3, both divided by $\sqrt{2}$. The rat race hybrid is the microstrip equivalent of the well-known waveguide magic tee.

Since the operation relies on the distance between ports being a multiple of $\lambda_g/4$, the performance is frequency dependent. Techniques for broadbanding were investigated [Agrawal and Mikucki 1986; Mikucki and Agrawal 1989].

6.5.4   Crossing

In the symmetrical crossing (Figure 6.19) one transmission line (input) is connected to three lines (outputs). If all the lines are identical, the approximate scattering matrix, neglecting reactive elements within the junction, is

$$
(\underline{s}) = \frac{1}{2}\begin{pmatrix} -1 & 1 & 1 & 1 \\ 1 & -1 & 1 & 1 \\ 1 & 1 & -1 & 1 \\ 1 & 1 & 1 & -1 \end{pmatrix}. \tag{6.38}
$$

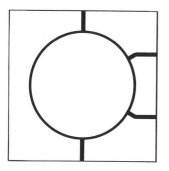

FIGURE 6.17   Rat race hybrid ring.

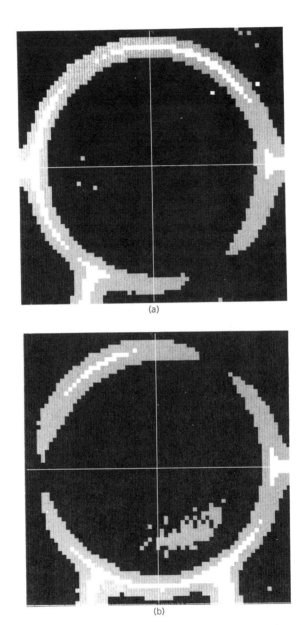

(a)

(b)

FIGURE 6.18 Distribution of signals on a rat race hybrid ring. (Courtesy J-F. Zürcher.)

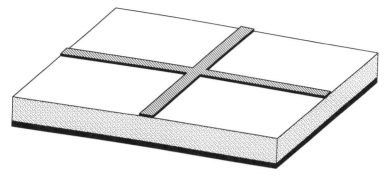

FIGURE 6.19   Crossing.

The input VSWR, with all output ports matched, is then 3 : 1. The crossing behaves like a two-dimensional TLM node (Section 10.4.4).

Crossings made with microstrip lines with different impedances were analyzed with the waveguide model [Menzel and Wolff 1977] and by integral techniques [Gopinath et al. 1976; Wu et al. 1990].

6.6   DIRECTIONAL COUPLERS

6.6.1   Coupled Microstrip Lines

Several microstrip devices such as directional couplers and filters (Section 7.4) use coupled lines, made of two strips that run parallel on top of the substrate. In most cases the two coupled lines have equal widths, and the resulting structure is symmetrical (Figure 6.20).

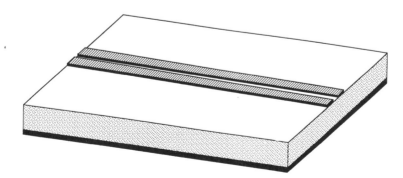

FIGURE 6.20   Symmetrical microstrip coupled line.

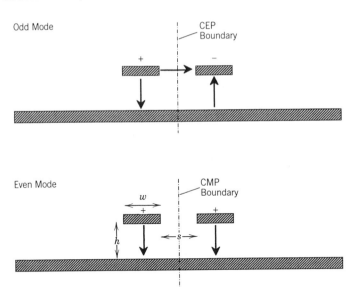

FIGURE 6.21 Even and odd modes on symmetrical coupled line.

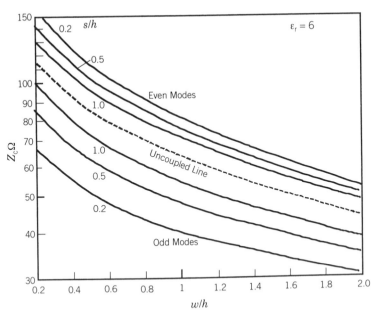

FIGURE 6.22 Characteristic impedances for the even and odd modes.

A signal propagating on a coupled line can be described by the superposition of an even mode and an odd mode (Figure 6.21) that exhibit different propagation factors and characteristic impedances (or admittances).

The techniques used to analyze microstrip transmission lines (Chapter 3) were extended to study coupled lines. Some of the main developments can be credited to Bryant and Weiss [1968]; Akhtarzad et al. [1975], while the results most often used nowadays are due to Kirschning and Jansen [1984]. The even- and odd-mode characteristic impedances for a symmetrical coupled line are shown in Figure 6.22 as a function of geometric parameters.

6.6.2 The Even Mode

In the even mode, the voltages and currents are identical on both strips. The electric field lines are tangential to the plane of symmetry, which can be replaced by an open-circuit or PMC boundary. Connecting a device across the two conductors does not affect the even mode because the two end points are at the same potential.

6.6.3 The Odd Mode

In the odd mode, the voltages and currents on both strips have the same amplitude but opposite directions. The electric field lines are normal to the plane of symmetry, which can be replaced by a short-circuit or PEC boundary. Since the two conductors are at different potentials, connecting a device between them affects the odd mode.

6.6.4 The Directional Coupler: A Symmetrical Four-Port

The developments made in Sections 6.4.5–6.4.7 can be applied to the analysis of uniform couplers (Figure 6.23).

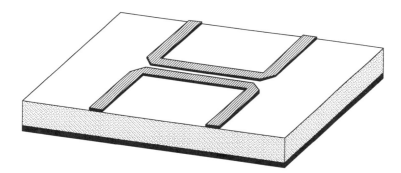

FIGURE 6.23 Uniform symmetrical coupler.

Symmetrical top–bottom excitations imply the existence of an open-circuited horizontal plane of symmetry. This corresponds to the even mode of the coupled line, with the corresponding input admittances:

$$\underline{Y}_{ss} = jY_e \tan \beta_e \frac{L}{2}, \qquad \underline{Y}_{as} = -jY_e \cot \beta_e \frac{L}{2}. \qquad (6.39)$$

Conversely, antisymmetrical excitations correspond to the odd mode, for which

$$\underline{Y}_{sa} = jY_o \tan \beta_o \frac{L}{2}, \qquad \underline{Y}_{aa} = -jY_o \cot \beta_o \frac{L}{2}. \qquad (6.40)$$

The four reflection factors are then given, in the input plane, by

$$\underline{\rho}_{ss} = \frac{Y_c - jY_e \tan(\beta_e L/2)}{Y_c + jY_e \tan(\beta_e L/2)}, \qquad \underline{\rho}_{as} = \frac{Y_c + jY_e \cot(\beta_e L/2)}{Y_c - jY_e \cot(\beta_e L/2)},$$

$$\underline{\rho}_{sa} = \frac{Y_c - jY_o \tan(\beta_o L/2)}{Y_c + jY_o \tan(\beta_o L/2)}, \qquad \underline{\rho}_{aa} = \frac{Y_c + jY_o \cot(\beta_o L/2)}{Y_c - jY_o \cot(\beta_o L/2)}. \qquad (6.41)$$

The scattering matrix is determined from Eqs. 6.18 and 6.25, which are regrouped here for convenience:

$$\underline{S}_{11} = \underline{S}_{22} = \underline{S}_{33} = \underline{S}_{44} = \underline{S}_1 = \underline{\rho}_{ss} + \underline{\rho}_{as} + \underline{\rho}_{sa} + \underline{\rho}_{aa},$$

$$\underline{S}_{12} = \underline{S}_{21} = \underline{S}_{34} = \underline{S}_{43} = \underline{S}_2 = \underline{\rho}_{ss} - \underline{\rho}_{as} + \underline{\rho}_{sa} - \underline{\rho}_{aa},$$

$$\underline{S}_{13} = \underline{S}_{31} = \underline{S}_{24} = \underline{S}_{42} = \underline{S}_3 = \underline{\rho}_{ss} + \underline{\rho}_{as} - \underline{\rho}_{sa} - \underline{\rho}_{aa},$$

$$\underline{S}_{14} = \underline{S}_{41} = \underline{S}_{23} = \underline{S}_{32} = \underline{S}_4 = \underline{\rho}_{ss} - \underline{\rho}_{as} - \underline{\rho}_{sa} + \underline{\rho}_{aa}. \qquad (6.42)$$

The sums and differences of reflection factors for the even and odd modes yield

$$\underline{\rho}_{ss} + \underline{\rho}_{as} = \frac{2j(Y_c^2 - Y_e^2) \sin \beta_e L}{2Y_c Y_e \cos \beta_e L + j(Y_c^2 + Y_e^2) \sin \beta_e L},$$

$$\underline{\rho}_{ss} - \underline{\rho}_{as} = \frac{4Y_c Y_e}{2Y_c Y_e \cos \beta_e L + j(Y_c^2 + Y_e^2) \sin \beta_e L},$$

$$\underline{\rho}_{sa} + \underline{\rho}_{aa} = \frac{2j(Y_c^2 - Y_o^2) \sin \beta_o L}{2Y_c Y_o \cos \beta_o L + j(Y_c^2 + Y_o^2) \sin \beta_o L},$$

$$\underline{\rho}_{sa} - \underline{\rho}_{aa} = \frac{4Y_c Y_o}{2Y_c Y_o \cos \beta_o L + j(Y_c^2 + Y_o^2) \sin \beta_o L}. \qquad (6.43)$$

Introducing these expressions into Eq. 6.42 yields all the terms of the scattering matrix.

6.6.5   Conditions for Matching

For a matched device, $\underline{s}_{11} = 0$, and one must then have

$$\frac{\left(Y_c^2 - Y_e^2\right)\sin \beta_e L}{Y_c Y_e \cos \beta_e L + j\left(Y_c^2 + Y_e^2\right)\sin \beta_e L}$$

$$+ \frac{\left(Y_c^2 - Y_o^2\right)\sin \beta_o L}{Y_c Y_o \cos \beta_o L + j\left(Y_c^2 + Y_o^2\right)\sin \beta_o L} = 0. \qquad (6.44)$$

This expression provides the condition that must be met by the propagation factors, the characteristic admittances of the two modes, and the length of the coupled section.

6.6.6   Matched Directional Coupler

Equation (6.44) admits a frequency-independent solution when the odd and even modes of the line are degenerate; that is, when their propagation factors are identical

$$\beta_e = \beta_o = \beta, \qquad (6.45)$$

and when the characteristic admittances of the even and odd modes are selected so that

$$Y_o Y_e = Y_c^2. \qquad (6.46)$$

The coupler is matched ($\underline{s}_1 = 0$) and then $\underline{s}_4 = 0$, so when port 1 is the input port, port 4 is decoupled, instead of port 3 in the branch line coupler. The coupler is thus *contradirectional*. The other two terms are

$$\underline{s}_2 = \frac{2Y_c Y_e}{2Y_c Y_e \cos \beta L + j\left(Y_c^2 + Y_e^2\right)\sin \beta L},$$

$$\underline{s}_3 = \frac{j\left(Y_c^2 - Y_e^2\right)\sin \beta L}{2Y_c Y_e \cos \beta L + j\left(Y_c^2 + Y_e^2\right)\sin \beta L}. \qquad (6.47)$$

The resulting coupling and insertion loss are shown in Figure 6.24 as a function of βL with Y_e/Y_c as parameter.

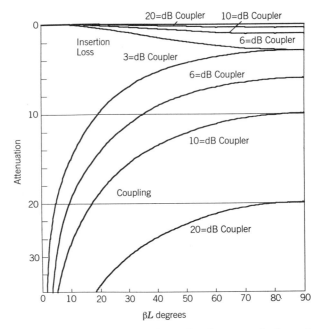

FIGURE 6.24 Insertion loss and coupling for a matched coupler.

6.6.7 Coupler on Inhomogeneous Microstrip

Within homogeneous coupled structures, in which the fields are contained within a single propagation medium, both the even and the odd modes are pure TEM, in which case the condition $\beta_e = \beta_o = \beta$ is automatically satisfied.

This is no longer true in inhomogeneous structures like microstrip, because the fields of the even mode are more concentrated within the dielectric than those of the odd mode, producing different propagation factors. This means that it is not possible to obtain a perfectly matched microstrip coupler, and some signal will also leak toward the isolated port.

Several schemes were proposed to reduce or correct this effect:

1. Multilayered structures [Haupt and Delfs 1974; Klein and Chang 1990].

2. Interdigital capacitors connected between the two strips. The capacitors do not affect the even mode (Section 6.6.2) but modify the propagation characteristics of the odd mode (Section 6.6.3) [Schaller 1977; Herzog 1978].

3. Combline couplers [Gunton 1978; Islam 1988].

6.6.8   Coupling, Isolation, and Directivity

Coupler parameters are most often specified in terms of the logarithmic attenuation levels (dB) defined in Section 4.3.9:

Coupling

$$LC = -20\log_{10}|\underline{s}_3|. \tag{6.48}$$

Insertion loss

$$LA = -20\log_{10}|\underline{s}_2|. \tag{6.49}$$

Isolation

$$LI = -20\log_{10}|\underline{s}_4|. \tag{6.50}$$

Directivity

$$LD = -20\log_{10}|\underline{s}_4/\underline{s}_3| = LI - LC. \tag{6.51}$$

The quality of a coupler is denoted by its directivity, which is the difference between the isolation and the coupling. It expresses the ratio of the unwanted signal in the isolated port to the coupled signal. Inhomogeneous couplers inherently possess a low directivity which can be improved to some extent by the schemes in Section 6.6.7.

6.6.9   Broad-Band Couplers

The coupling provided by a uniform section of coupled line is frequency dependent (Eq. 6.47). In actual couplers, one wishes to have a constant coupling over a specified frequency range that may be quite wide. Broadbanding of couplers is done either by cascading coupled line sections having different spacings [Levy 1963, 1964] or by using nonuniform coupled lines [Arndt 1968; Uysal and Watkins 1991].

6.6.10   Lange Coupler

The range of coupling that can be provided by couplers having two lines is rather limited, and additional coupled lines may be added when tighter couplings are required. The approach most often used is a four-line structure called the Lange coupler [Lange 1979; C. M. Jackson 1989; Waugh 1989] (Figure 6.25). This structure requires wire crossovers joining nonadjacent strips, so the device is not completely planar. This requirement produces an additional complication in the practical realization of the device, which may prove bothersome when large production is involved. The crossovers can now

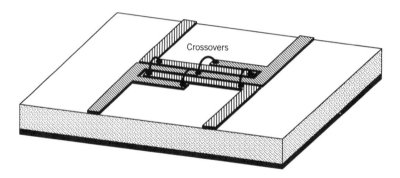

FIGURE 6.25 Lange coupler.

be realized with multilayer thick-film technology [Bourreau et al. 1992]. Coupling can also be increased by cascading simple directional couplers.

6.7 HIGHER-ORDER MULTIPORTS

6.7.1 Five-Port Junctions

The scattering matrix of a symmetrical reciprocal matched five-port (Figure 6.26) contains only two independent parameters, the transmission factors to adjacent ($\underline{A}$) and to nonadjacent ($\underline{B}$) ports:

$$(\underline{s}) = \begin{pmatrix} 0 & \underline{A} & \underline{B} & \underline{B} & \underline{A} \\ \underline{A} & 0 & \underline{A} & \underline{B} & \underline{B} \\ \underline{B} & \underline{A} & 0 & \underline{A} & \underline{B} \\ \underline{B} & \underline{B} & \underline{A} & 0 & \underline{A} \\ \underline{A} & \underline{B} & \underline{B} & \underline{A} & 0 \end{pmatrix}. \tag{6.52}$$

When the device is lossless (Eq. 4.25), three independent equations result:

$$2|\underline{A}|^2 + 2|\underline{B}|^2 = 1,$$

$$|\underline{A}|^2 + \underline{A}^*\underline{B} + \underline{A}\underline{B}^* = 0,$$

$$|\underline{B}|^2 + \underline{A}^*\underline{B} + \underline{A}\underline{B}^* = 0. \tag{6.53}$$

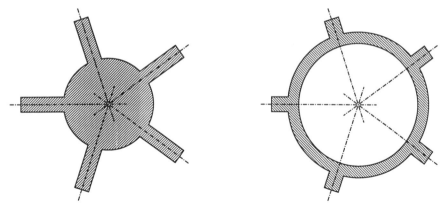

FIGURE 6.26   Symmetrical five-ports.

These conditions are satisfied when

$$|\underline{A}| = |\underline{B}| = 1/2 \quad \text{and} \quad \arg(\underline{A}) - \arg(\underline{B}) = 2\pi/3. \quad (6.54)$$

A signal fed to one port is evenly distributed among the other four ports (6-dB insertion loss). A matched lossless symmetrical five-port junction is thus a four-way power divider. Several authors considered practical implementations [Gupta and Abouzhara 1985; Yeo et al. 1989; Wang et al. 1991].

6.7.2 Six-Port Junctions

The scattering matrix of a symmetrical reciprocal six-port junction (Figure 6.27) can be expressed as follows [Yeo 1992]:

$$(\underline{S}) = \begin{pmatrix} \gamma & \underline{\alpha} & \underline{\beta} & \underline{\tau} & \underline{\beta} & \underline{\alpha} \\ \underline{\alpha} & \gamma & \underline{\alpha} & \underline{\beta} & \underline{\tau} & \underline{\beta} \\ \underline{\beta} & \underline{\alpha} & \gamma & \underline{\alpha} & \underline{\beta} & \underline{\tau} \\ \underline{\tau} & \underline{\beta} & \underline{\alpha} & \gamma & \underline{\alpha} & \underline{\beta} \\ \underline{\beta} & \underline{\tau} & \underline{\beta} & \underline{\alpha} & \gamma & \underline{\alpha} \\ \underline{\alpha} & \underline{\beta} & \underline{\tau} & \underline{\beta} & \underline{\alpha} & \gamma \end{pmatrix}. \quad (6.55)$$

For an ideal matched lossless junction, $\gamma = \beta = 0$, $\underline{\alpha} = (1/\sqrt{3})\exp(j\phi_\alpha)$, $\underline{\tau} = (1/\sqrt{3})\exp(j\phi_\tau)$ with $\phi_\alpha - \phi_\tau = 2\pi/3$. The input port is coupled to the

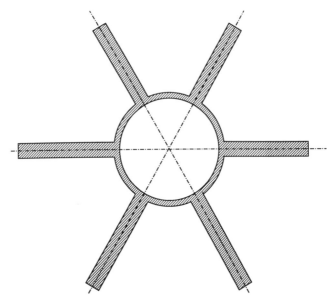

FIGURE 6.27 Symmetrical six-port.

two adjacent and to the opposite ports, while the remaining two ports are isolated.

6.7.3 Radial Line Dividers

Another class of power dividers was proposed by Abouzhara and Gupta [1988], in which the input line is coupled to the narrow end of a radial line sector (Figure 6.28). At the other end, an arbitrary number of output lines can be connected to the wide end of the radial line sector. These devices can

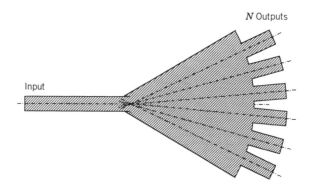

FIGURE 6.28 Sector-shaped components.

provide an approximately equal split of the input signal among the output ports, but the latter cannot be matched.

6.8 PROBLEMS

6.8.1   Analyze a lossless reciprocal three-port that is matched at two of its three ports.

6.8.2   One wishes to realize a power divider with a power ratio of 2.5 : 1 between the two outputs. Determine the characteristic impedances of the two output lines, with a 50-Ω input, and evaluate all the terms of the scattering matrix.

6.8.3   The design of the matched three-port resistive power divider can be generalized for n-port resistive dividers. Find the value of the resistors required and the terms of the resulting scattering matrix.

6.8.4   Show that by connecting two ideal three-port circulators, one obtains an ideal four-port circulator.

6.8.5   One wishes to realize a branch line coupler with a power ratio of 2.5 : 1 between the two coupled outputs. Determine the characteristic impedances of the two lines in the coupler, in the case of a 50-Ω input, and evaluate all terms of the scattering matrix.

6.8.6   Determine the scattering matrix of a crossing between two lines having characteristic impedances of 50 Ω and 100 Ω.

6.8.7   One wishes to realize a single-section 18-dB coupler having the largest possible frequency bandwidth. Determine the even and odd characteristic admittances of the coupled line.

6.8.8   Three-port and four-port power dividers can be realized with a symmetrical five-port matched junction in which two ports, respectively one port, are terminated into matched loads. Determine the scattering matrices of the two devices and compare them with those of resistive three- and four-port power dividers (Section 6.2.4, problem 6.8.3).

6.8.9   Determine the scattering matrix of a symmetrical line divider with five outputs. The device is reciprocal, lossless, and matched at its input, dividing the input power equally (amplitude and phase) between the five outputs.

CHAPTER SEVEN

Resonators and Filters

7.1 PATCH RESONATORS

7.1.1 General Description

An upper conductor of finite dimensions, or a patch, is deposited on the dielectric substrate (Figure 7.1). Patches may take several shapes, depending on the applications: rectangular, circular, triangular, and so on. The amplitude of the currents that flow on the patch becomes significant when the signal frequency is close to a resonance (eigenvalue). The current patterns at resonance correspond to modes, or eigenfunctions, of the structure. Resonances start to occur when the patch size is about half-wavelength.

When a resonant patch is loosely coupled to transmission lines (ports), transmission across the patch occurs near the resonant frequencies, and a resonant patch behaves like a simple band-pass filter. Higher-order filters are realized by combining several resonators (Sections 7.3 and 7.4). When the coupling of the patch to the connecting lines becomes large, the resonant character vanishes and one obtains a junction (Chapter 6).

Patch resonators can also radiate and be used to realize antennas (Chapter 11). This is clearly unwanted for circuit elements. Excessive radiation is avoided by properly choosing the substrate (Section 1.3.6).

7.1.2 Rectangular Patch

A rectangular patch is a section of transmission line of length L. The fringing fields at the open ends are taken into account by adding equivalent lengths ΔL at both ends (Eqs. 5.1 or 5.2). The resonant frequencies are then given by

$$ f_m = \frac{mc_0}{2(L + 2\Delta L)\sqrt{\varepsilon_{\text{ef}}}}, \qquad (7.1) $$

123

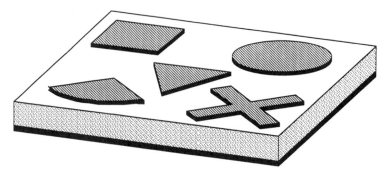

FIGURE 7.1 Patch resonators.

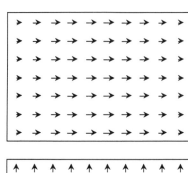

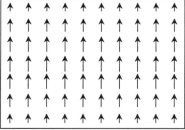

FIGURE 7.2 Current distributions of resonator modes on a rectangular patch.

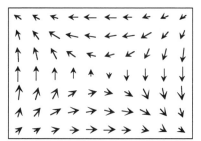

where $m = 1, 2, 3$ is an integer, and where the relative permittivity ε_{ef} is given by Eqs. 3.39 or 3.42.

This expression is valid for narrow lines that cannot resonate transversely. Additional resonances appear on wide lines, corresponding to the higher-order modes of the microstrip (Section 3.7.1). The current distributions for several modes on a rectangular patch are shown in Figure 7.2.

7.1.3 Circular Patch

Circular resonator patches lose less energy by radiation, and thus provide larger quality factors than rectangular ones [Watkins 1969]. The resonant frequency is determined by assuming that a perfect magnetic wall (PMC, Section 2.3.6) extends under the edges of the patch (Figure 7.3). Fringing fields are taken into account by defining an effective resonator radius a_{eff}, slightly larger than the physical radius a.

In practical situations, the substrate height is much smaller than the radius, and the first modes of resonance are transverse magnetic TM_{mn0} modes, with resonant frequencies

$$f_{mn} = \frac{c_0 \chi_{mn}}{2\pi a_{\text{eff}} \sqrt{\varepsilon_{\text{r}}}}, \tag{7.2}$$

where χ_{mn} is the mth extremum of the Bessel function J_n. The effective radius was determined by Itoh and Mittra [1973]:

$$a_{\text{eff}} = a\sqrt{1 + (2h/\pi a) \ln(\pi a/2h + 1.7726)}. \tag{7.3}$$

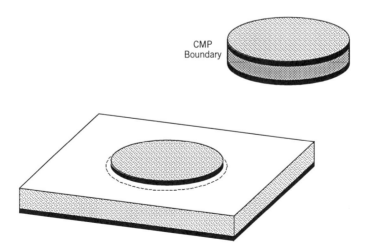

FIGURE 7.3   Disk resonator and equivalent model.

The unloaded quality factor for the fundamental TM_{110} mode was evaluated by Dydyk [1986] as

$$Q_0 = \frac{1}{\lambda R_m / \pi Z_0 h + \tan \delta},$$

(7.4)

where λ is the free-space wavelength, R_m is the metal resistance (Eq. 2.43), and $Z_0 = 120\pi$ Ω is the free-space impedance. This expression includes metal losses and dielectric losses but does not take into account radiation.

7.1.4 Ring Resonators

Ring resonators are formed by bending a microstrip line which is several wavelengths long (Figure 7.4). End effects are then avoided (fringing fields and radiation). The effect of curvature was determined by several authors [Owens 1976; Xu and Bosisio 1992], but it is not significant when the line is very thin ($w \ll R$). Resonances occur for

$$\lambda_g = 2\pi R / n,$$

(7.5)

where R is the median radius and n is an integer.

Ring resonators provide a simple experimental means to determine the wavelength λ_g and are used to characterize microstrip lines [Wolff and Knoppik 1971; Wu and Rosenbaum 1973], discontinuities [Hoefer and Chattopadhyay 1975], and substrates (Section 14.2). They are also encountered in the design of filters [Guglielmi and Fernandez 1990] and tunable oscillators [Chang et al. 1987].

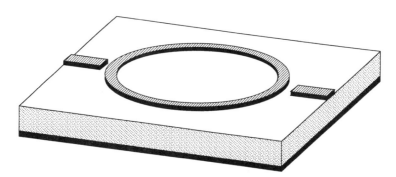

FIGURE 7.4 Microstrip ring resonator.

7.1.5 Other Shapes

The resonant frequencies of patches having other shapes are more difficult to determine (Figure 7.1). Equivalent cavity methods of Section 10.3 are used for their analysis, while integral methods (Section 10.6) evaluate radiation and surface-wave effects.

7.1.6 Tunable Resonators

The resonant frequency of a microstrip resonator can be adjusted over a limited range by electronic means, namely by incorporating a reverse-biased semiconductor diode (varactor, Section 9.16) in the resonator circuit. A biasing voltage must be provided for the varactor [Howes and Morgan 1978].

7.2 DIELECTRIC RESONATORS

7.2.1 High-Permittivity Dielectrics

Low-loss ceramics with very high permittivities are now available (Section 13.1.6). They can be used to reduce the size of resonators and filters.

Materials with relative permittivities $\varepsilon_r = 38$ and $\varepsilon_r = 88$ were described by Tamura et al. [1988]. Ceramics with permittivities $\varepsilon_r = 21.6-152$ are also available (DI-MIC, Dielectric Laboratories Inc.).

7.2.2 High-Permittivity Resonators

Resonators made with high-permittivity ceramics are most often of circular cylindrical (pillbox) shape. They are easily inserted into printed circuits next to a microstrip line, with coupling provided by the fringing fields of the line (Figure 7.5).

7.2.3 Resonant Frequencies

Since the permittivity of the resonator is much larger than that of the surrounding air, the resonator walls are replaced, in first approximation, by perfect magnetic conductors (open circuits, PMC, Section 2.3.6). The theoretical analysis is then the dual of the one of metallic cavities [Gardiol 1984]. Resonant frequencies for a resonator of radius a and of height d are

$$f_{mnl} = \frac{c_0}{2\sqrt{\varepsilon_r}} \sqrt{\left(\frac{\chi_{mn}}{\pi a}\right)^2 + \left(\frac{l}{d}\right)^2}, \tag{7.6}$$

where χ_{mn} is the mth extremum of the Bessel function J_n for a TM mode, or

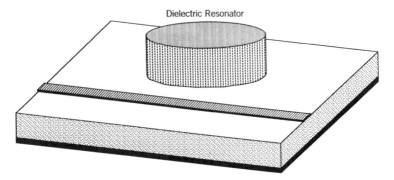

FIGURE 7.5   Dielectric resonator coupled to a transmission line.

its mth zero for a TE mode. The integer l denotes the number of half-wavelengths along the vertical dimension. The situation is similar to the one encountered in Section 7.1.3, but with magnetic conductor boundaries on the top and bottom (instead of the metal walls in Figure 7.3).

The fringing fields can be taken into account by a perturbation method [Gardiol 1984]. A cylindrical resonator may also be considered to be formed by a section of a dielectric waveguide (optical fiber), terminated by open circuits at both ends. When the line coupling and the presence of a substrate are also considered, the study becomes complex and requires a computer [Van Bladel 1975; Murray 1989; Mongia 1990; Hearn et al. 1990].

7.3   LOW-PASS FILTERS

7.3.1   Description

A low-pass filter transmits low-frequency signals but reflects signals above an upper bound called the cutoff frequency. Low-pass filters isolate the bias supply from the microwave circuit, keeping high-frequency signals away from dc voltage or current supplies. They also separate the intermediate-frequency components obtained at the mixer's output in a receiver.

7.3.2   Principle of Operation

Low-pass filters are realized by cascading a transmission line with an alternating series of wide strips (low impedance) and narrow strips (high impedance). The steps produce reflections (Section 5.3), which add up vectorially, with phase shifts related to the section lengths. The reflections tend to cancel each other when frequencies are smaller than the filter's cutoff frequency. On the other hand, they build up, yielding almost complete reflection, for frequencies beyond the cutoff (creating stop bands).

The range of strip widths that can actually be realized is limited. Narrow strips are difficult to manufacture accurately, due to uncertainties in the etching process (undercutting, Section 13.3.8). On the other hand, when the strip becomes large, transverse resonances may appear. The minimum and maximum characteristic impedances are thus key parameters in the design of low-pass filters.

7.3.3   Filter Synthesis

The design of a microstrip filter starts by using well-known techniques developed for circuit theory synthesis [Matthaei et al. 1964; Malherbe 1979]. A lumped-element ladder network with series inductors and shunt capacitors satisfying the required specifications is first synthesized. Two kinds of schemes can be used for this purpose:

1. In a Butterworth filter, all the attenuation zeros are placed at the zero frequency, providing a maximally flat attenuation within the passband. Attenuation increases monotonically with frequency, reaching 3 dB at the cutoff.
2. In a Chebyshev filter, the attenuation zeros are evenly distributed within the passband, producing a ripple.

The degree of the filter required to meet the specifications is determined. The inductances and capacitances are then realized with sections of microstrip lines, wide sections being mostly capacitive, narrow ones mostly inductive (Section 4.6.3).

The synthesis process takes into account the parasitic reactances at the discontinuities provided by the equivalent circuit of the step (Section 5.3). Successive steps can be rather close to each other, so coupling between steps must also be taken into account.

7.3.4   Example of Realization

The layout software Micros (Section 12.4.5) includes the synthesis of low-pass filters (Chebyshev or Butterworth) with up to 15 stages and carries out the design in a few seconds, taking the coupling between successive steps into account.

The desired filter must have its cutoff frequency at 3.5 GHz and an attenuation of at least 35 dB at 6 GHz. The substrate selected has a relative permittivity of $\varepsilon_r = 10.7$ and a thickness $h = 0.635$ mm. The computer determines the line widths: $w_0 = 0.557$ mm for the end connections (50 Ω), $w_b = 3.69$ mm for the broad strip, and $w_n = 0.0856$ mm for the narrow strip. The lengths of the successive sections are 1.786, 2.675, 2.7, and 2.918 mm

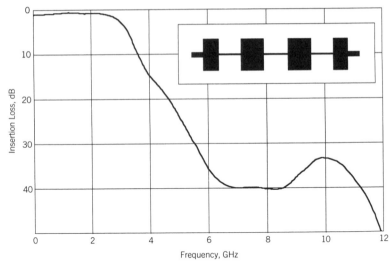

FIGURE 7.6   Low-pass filter.

(center section), the structure being symmetrical. The design was realized, and the measured attenuation is represented in Figure 7.6.

7.4   BAND-PASS FILTERS

7.4.1   Description

A band-pass filter transmits signals within a specified frequency band and reflects signals outside of this band. Band-pass filters select a part of the frequency spectrum and the signals contained therein. They permit one to separate useful signals from unwanted signals and noise.

7.4.2   Filter Synthesis

A band-pass filter is derived from a low-pass filter by frequency shifting. In the ladder network for the low-pass filter (Section 7.3.3), series inductors are replaced by antiresonant series cells, shunt capacitors by resonator shunt cells. In the Butterworth approach, the attenuation zeros are placed at the center of the frequency band, while in a Chebyshev filter they are distributed evenly within the passband.

The degree of the filter required to meet the specifications is first determined. The series and shunt cells are then realized with sections of transmission lines and coupled transmission lines. The synthesis process must take into account the parasitic reactances at all the discontinuities, including those at the ends of open stubs (Chapter 5).

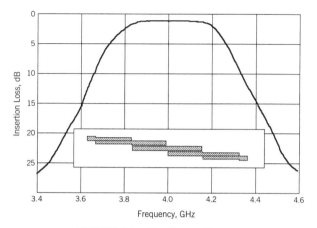

FIGURE 7.7   Band-pass filter.

7.4.3   Example of Realization

We wish to realize a band-pass filter meeting the following requirements with a passband 3.8–4.2 GHz:

passband ripple	0.3 dB
stop-band attenuation	25.0 dB at 4.6 GHz
substrate permittivity	2.33 (relative)
substrate thickness	0.51 mm

These requirements are introduced into a CAD program for the synthesis of filters (Section 7.5). For instance, the layout package Micros (Section 12.4.5) carries out the complete synthesis process. It determines the response for a Chebyshev design, the order of the filter, calculates the dimensions of the strips and the spacings of coupled lines, and draws the layout with broadside-coupled resonator strips. A band-pass filter is designed within about 2 min.

The filter designed was fabricated and measured on a vector network analyzer (Section 14.1). The measured transmission factor is shown in Figure 7.7. The passband falls on the specified range, and the 25-dB off-band requirement is also met [Zürcher and Gardiol 1989].

7.4.4   Hairpin Resonators

The resonator strips in Figure 7.7 are approximately $\lambda_g/2$ long, so the surface required for the filter can be a significant portion of the whole surface of the circuit. The strips are actually dipole antennas, which may radiate [Schafer et al. 1989]. The size can be reduced and the radiation

FIGURE 7.8   Hairpin resonators.

avoided by bending the strips to realize hairpin resonators [Cristal and Frankel 1972]. The effect of the bends must be taken into account (Section 5.4). Further reduction is possible by bending over the ends of the hairpins and completing the design with a section of coupled line [Sagawa et al. 1989; Roduit et al. 1991] (see Figure 7.8).

7.5   FILTER SOFTWARE

Some of the software packages listed in Sections 12.3 and 12.4 include programs for the analysis and design of filters, while in some cases this task is performed by a compatible companion program [Denig 1989]. Moderately complex filters can be designed accurately on a desktop computer [Railton and Meade 1992].

7.5.1   Planim

The planar circuit approach for the characterization of microstrip circuits is combined with the image parameter method, giving a powerful technique for the design of microstrip filters [Salerno and Sorrentino 1986]. The inclusion of two-dimensional effects in the synthesis procedure makes the planim approach particularly suited for monolithic microwave circuits.

7.5.2 Scallop

This software converts a lumped-element Chebyshev band-pass filter, designed with classical methods, into a microstrip filter. The use of asymmetrical lines provides additional degrees of freedom. Design can be done with
either shorted or open-ended stubs, depending on the available technology
(ArguMens, Section 12.3.6).

7.6 PROBLEMS

7.6.1 A rectangular patch 15 mm long and 1.1295 mm wide is deposited on
top of a substrate 0.5 mm thick having a relative permittivity $\varepsilon_r = 3.5$.
Determine the resonant frequencies.

7.6.2 A circular patch of radius $a = 4$ mm is deposited on a substrate of
thickness $h = 0.6$ mm and relative permittivity $\varepsilon_r = 6.5$. Determine
the resonant frequency of the mode TM_{110} (for which $\chi_{11} = 1.841$).

7.6.3 A ring resonator having a radius $R = 20$ mm resonates at 3.27 GHz.
Determine the effective permittivity of the substrate.

7.6.4 Determine approximately the radius of a cylindrical dielectric resonator of relative permittivity $\varepsilon_r = 80$ to make it resonate on the
TM_{110} mode at 7.6 GHz.

7.6.5 Find the frequency dependence of three cascaded sections of transmission lines $\lambda_g/4$ long at 3 GHz, the characteristic impedance of the
center section being 250 Ω and those of the two adjacent sections 10
Ω. The assembly is connected at both ends to 50 Ω lines.

7.6.6 A duplexer (frequency-selective divider) is made by connecting two
band-pass filters with different passbands at the two outputs of a
symmetrical power divider (Section 6.2.1). Determine the input
impedance of filter 2 within the passband of filter 1 that is required to
ensure full transfer of the input power across filter 1.

CHAPTER EIGHT

Transistor Amplifiers and Oscillators

The discovery of the transistor effect in 1948 revolutionized the field of electronics, previously based on the use of bulky and power-consuming electron tubes. Transistors quickly led to the development of sophisticated data processing, opening the whole era of computer electronics. Just as printed circuits for low-frequency electronics were developed to assemble transistor circuits, the development of microstrips is closely related to the increasing availability of transistors operating at microwave frequencies.

8.1 GENERAL DESCRIPTION

8.1.1 Outline of Transistor Circuits

Transistors are three-terminal semiconductor devices that, with associated circuitry, provide amplification over certain frequency ranges. Two main techniques were developed to realize transistors, which can be either bipolar (Section 8.2) or field effect (Section 8.3). One of the three terminals is generally grounded, and the two remaining ones are connected to matching and biasing networks. The general block diagram of a microwave transistor amplifier is shown in Figure 8.1.

The matching circuits are required on both sides to adapt the input and output impedances of the transistor to the characteristic impedance of connecting transmission lines, which at microwaves is usually normalized at 50 Ω. The power amplification provided by the transistor is given by

$$\frac{P_{out}}{P_{in}} = \left(\frac{i_{out}}{i_{in}} \right)^2 \frac{\mathrm{Re}(\underline{Z}_{out})}{\mathrm{Re}(\underline{Z}_{in})}, \tag{8.1}$$

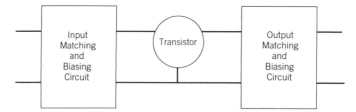

FIGURE 8.1   Block diagram of a transistor amplifier.

where the subscripts out and in stand for output and input, respectively, and the ratio i_{out}/i_{in} is the transistor's current gain. The power gain depends not only on the current gain but also on the impedance levels at input and output.

8.1.2 Frequency Limitations

Transistors were originally used in low-frequency applications because their operation is inherently frequency limited by two physical effects:

1. It takes a finite time τ for a charge carrier (electron or hole) to cross a junction and travel over the distance separating the electrodes connected to the semiconductor. Furthermore, charge carriers travel over paths that have different lengths, so a signal tends to spread out in time (dispersion). When the transit time becomes comparable to the period of the signal, the amplification provided by the transistor drops below unity and the device no longer amplifies.
2. Electrodes deposited over a semiconductor which has a significant permittivity form a capacitor. Reactive currents can thus flow across the device, competing with the conduction current provided by the charge carriers. In addition, the connecting wires contribute inductances. The combined effects of the reactive elements tend to yield a low-pass filter response.

Due to the combined effects of transit time and of reactive elements, the current gain of the transistor i_{out}/i_{in} decreases with frequency as $1/f$, becoming unity at the cutoff frequency f_c. The power gain thus decreases with the square of frequency as

$$\frac{P_{out}}{P_{in}} = \left(\frac{f_c}{f}\right)^2 \frac{\mathrm{Re}(\underline{Z}_{out})}{\mathrm{Re}(\underline{Z}_{in})}. \tag{8.2}$$

To have a sufficient power amplification at microwaves, one should have a high cutoff frequency f_c, together with a low input impedance $\underline{Z}_{in}$ and a large output impedance $\underline{Z}_{out}$. The device geometry and the doping levels must be optimized to provide the best possible compromise.

To increase the frequency of operation, the transit time must be decreased; that is, the device dimensions must be reduced. However, reducing the size also limits the current-carrying capability and decreases the gain. A reduction of the interelectrode spacing also tends to increase the capacitance. Technological advances are overcoming these limitations, and the operating frequency range of transistors now reaches millimeter waves.

8.1.3 Microwave Transistors

Transistors are currently available for operation at microwave frequencies, following spectacular developments in solid-state technology. Compound III-V semiconductor materials like gallium arsenide (GaAs) provide larger carrier mobilities than silicon. The use of heterojunctions provides channels in which charge carriers can move even faster. Very accurate positioning of the masks during the manufacturing process brought down electrode dimensions into the submicrometer region. As a result of these advances, transit times of picoseconds can now be achieved.

The effects of parasitic reactances can be reduced by the careful selection of geometries and of material properties, while matching elements can be realized directly on the semiconductor substrate.

Bipolar transistors (Section 8.2) are generally used up to 5–6 GHz, while applications at higher frequencies (up to 20–30 GHz) are based on field-effect transistors (MESFETS and HEMTS, Section 8.3). The useful range of transistor operation actually reaches 100 GHz [Smith and Swanson 1989].

8.1.4 Packaging and Insertion into Microstrip Circuits

Transistors can be directly mounted on microstrips, which provide a firm mechanical base and metal strips for connections (printed circuits were actually developed to more easily connect consistently complex transistor circuits and to eliminate tedious wiring operations). The transistor is attached with epoxy, and the metal connections are made by bonding or soldering (Section 13.6).

Cases that degrade the performances of the chip as little as possible have been designed. The usual cases permit an easy insertion with minimal discontinuities (Figure 8.2). By directly mounting the transistors in chip form on the microstrip boards, the reactive effects due to packages can be avoided and performances improved.

Power transistors provide output powers between 10 and 100 W, depending on frequency [Shih and Kuno 1989]. Because they dissipate heat, holes must be drilled through the substrate to insert the device and provide adequate heat transfer to a solid metal ground. This may cause some difficulty when microstrips are realized on ceramic substrates (Section 13.6.7).

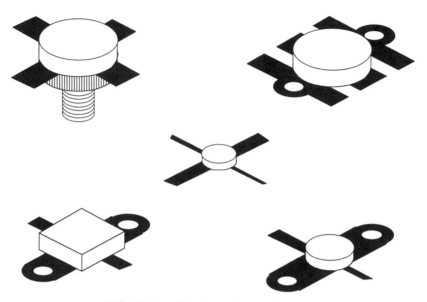

FIGURE 8.2 Usual transistor packages.

8.1.5 Remark

The field of microwave transistors is broad, and many technical articles and books have been devoted to it in recent years [Pengelly 1982; Soares 1988; Chang 1990; Yngvesson 1991; Ali and Gupta 1991; Golio 1991]. Research and development are quite active, extending the limits in power and frequency while reducing noise. Significant improvements can be expected within the next few years, as devices leave the development stage and enter the commercial field [Liechti 1989]. The next sections briefly describe the main operating principles used and the devices most likely to be encountered in microstrip circuits.

8.2 MICROWAVE BIPOLAR TRANSISTORS

8.2.1 Principle of Operation

Bipolar junction transistors reached the microwave range in the 1960s. They are generally realized in planar *npn* silicon technology and are currently used up to several gigahertz. Electrons are injected into the semiconductor by the emitter electrode and cross the emitter–base junction, which is polarized in the forward direction (Figure 8.3). Once in the base region, the electrons keep moving and cross the inversely polarized base–collector junction. A large negative bias is applied across the base–collector junction, while only a

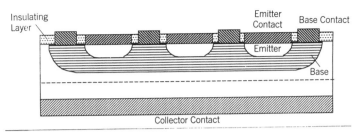

FIGURE 8.3 Bipolar transistor cross section.

small voltage is required across the emitter–base junction. Since the current flowing across the two junctions is almost the same, a signal applied at the input is amplified; that is, more power can be extracted on the collector side (output) than was supplied to the emitter–base side (input).

Bipolar transistors are made with a single-crystal n-doped silicon substrate (collector) on which an epitaxial layer of high-resistivity n-type silicon is generally grown. A p-doped region is then diffused or implanted on top of the silicon through a mask defining the base region, and, similarly, an n-doped emitter region is then diffused or implanted on top. The procedure is completed by the deposition of metal electrodes and insulating layers. Passivation surface treatments are required to avoid current leakage along the edges.

When a large gain is desired, the emitter is grounded and the base contact is used as input (common-emitter configuration), whereas more power is obtained by grounding the base and feeding the input signal to the emitter (common-base configuration).

8.2.2 Frequency Limitations

The cutoff frequency is inversely proportional to the transit time τ, which results from four contributions [Sze 1981; Yngvesson 1991]:

1. Charging time of the emitter–base junction
2. Transit time across the base
3. Transit time of the depletion layer of the base–collector junction
4. Charging time of the base–collector junction

Several geometries are utilized to widen the junctions as much as possible without increasing the surface of the electrodes (Figure 8.4). Metal–oxide capacitors and discrete components placed within the encapsulation allow matching to be performed over a wide frequency range.

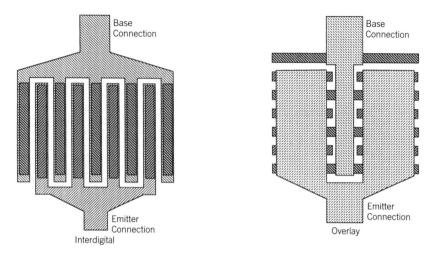

FIGURE 8.4   Bipolar transistor layouts.

8.2.3   Equivalent Circuit

The operation of a bipolar transistor can be simulated by the equivalent circuit of Figure 8.5, which includes the interelectrode capacitors, the lead inductances, and the substrate resistance. Many components in the circuit are actually nonlinear, so a complete simulation is quite complex. For the design of low-power amplifiers, a linear simulation proves generally adequate.

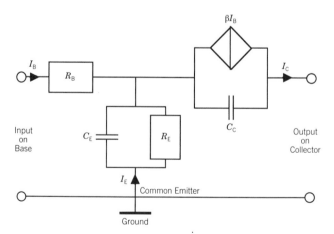

FIGURE 8.5   Equivalent circuit of bipolar transistor.

8.2.4   Heterojunction Bipolar Transistor

The frequency range of bipolar junction transistors was extended to 50–60 GHz by making use of an emitter–base heterojunction. Heterojunction bipolar transistors (HBTs) are realized in gallium arsenide technology (GaAs) with a GaAlAs emitter [Ali and Gupta 1991; Yngvesson 1991].

8.3   MICROWAVE FIELD-EFFECT TRANSISTORS

8.3.1   Basic Principle

In field-effect transistors (FETs) charge carriers travel along a channel in which the current flow can be controlled by a voltage bias. Carriers are injected by an electrode called the source and travel toward an electrode called the drain, with the gate electrode controlling the flow. The principle of operation is somewhat similar to that of a water faucet, in which the cross section of the channel can be mechanically changed to adjust the flow of liquid.

In contrast with bipolar transistors, carriers do not cross *pn* junctions but remain within the same type of semiconductor during the entire distance from source to drain. The transit time of carriers only includes the transit from source to drain since there are no junctions that must be charged. At microwaves, FETs are generally realized with III-V semiconductor materials, most often GaAs, in which the mobility of conduction electrons is six times larger than that in silicon, and the peak drift velocity is twice as large [Pengelly 1982].

8.3.2   MESFETS

In a metal–semiconductor field-effect transistor (MESFET) the channel is made of low-resistivity semiconductor material, generally *n*-doped GaAs deposited by epitaxy [Brodie and Muray 1992] on top of an undoped buffer layer, itself deposited over a semi-insulating (very high resistivity) substrate (Figure 8.6). The source and drain electrodes form ohmic metallic contacts, generally alloyed on highly doped n^+ GaAs pads. The gate electrode, placed in between, forms a rectifying Schottky junction (Section 9.1.2). A depletion layer extends below the junction, reducing the width of the channel [Pengelly 1982; Soares 1988]. The gate bias controls the width of the depletion layer and, thus, the electron current that can flow from the source to the drain electrodes. Reverse-biasing the gate junction allows a large source-to-drain current flow to be controlled by a small reverse gate current, providing current amplification. The operation is similar to that of an electronic triode.

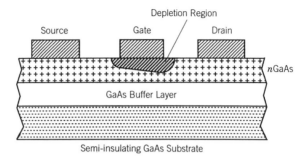

FIGURE 8.6 Cross section of a MESFET.

Increasing the negative gate bias up to the pinch-off voltage completely blocks current flow in the channel.

8.3.3 HEMTS

More recently developed high electron mobility transistors (HEMTs) use heterojunctions, that is, junctions between semiconductor materials that have different energy gaps (Figure 8.7).

The difference in conduction-band energies of the two materials produces a potential well on one side of the heterojunction in which the electrons concentrate and form a two-dimensional electron gas. They travel in un-doped material, physically separated from the donor ions, so that impurity scattering is reduced and mobility increased. The overall performance is therefore superior to that of MESFETs. The level of the potential well is adjusted by the gate bias, which thus controls the source-to-drain current.

HEMTs are variously known as modulation-doped FETs (MODFETs), two-dimensional electron-gas FETs (TEGFETs), selectively doped heterostructure FETs (SDHT), or heterostructure FETs (HFET). Some devices can operate at frequencies up to 100 GHz [Smith and Swanson 1989].

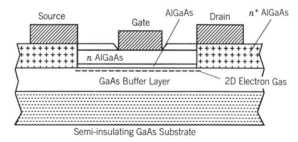

FIGURE 8.7 Cross section of a HEMT.

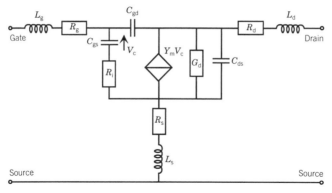

FIGURE 8.8   Equivalent circuit.

8.3.4   Equivalent Circuit

Even though the physical principles of operation are somewhat different for MESFETS and HEMTS, their topology is quite similar, and they can be described by the same equivalent circuit (Figure 8.8).

Considerable effort was carried out to provide accurate transistor models, in particular to improve the design of MMICs [Salmer et al. 1988]. A correspondence between equivalent circuits and matrix representations was given by Twisleton [1991].

Reactive internal feedback from drain to source creates low-frequency instabilities: MESFETS are inherently microwave devices, whose operation at frequencies lower than several gigahertz is not recommended [Ohkawa et al. 1975]. During transistor circuit design, precautions must be taken to avoid spurious oscillations at lower frequencies (Section 8.4.3).

The cutoff frequency f_c can be derived from the equivalent circuit and is inversely proportional to the transit time τ:

$$f_c = \frac{g_m}{2\pi C_{gs}} = \frac{1}{2\pi\tau} \cong \frac{v_s}{2\pi L_g}, \tag{8.3}$$

where L_g is the effective length of the gate, and it is assumed that electrons travel at their saturation velocity v_s [Yngvesson 1991].

8.3.5   Electrode Geometry

The cutoff frequency is inversely proportional to the gate length L_g (Eq. 8.3). The high-frequency performance of FETs depends critically on the gate length, and sophisticated technological schemes were developed to reduce this length to the smallest possible dimensions.

In the self-aligned gate technology, gate metal is deposited in a first step, covered with photoresist, and the desired source and drain pattern realized by the photolithographic process. The exposed gate metal is then removed by etching, undercutting being used to define the spacings between the gate and the other two electrodes. Ohmic metal is then deposited over the structure,

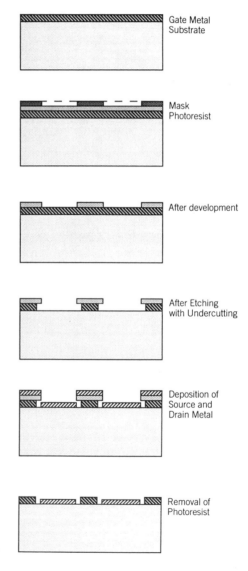

FIGURE 8.9   Self-aligned process.

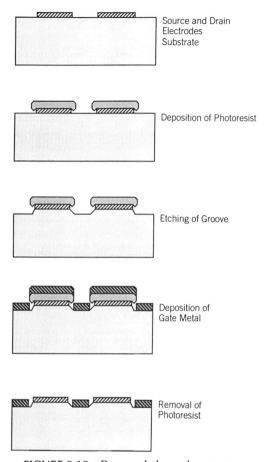

FIGURE 8.10   Recessed channel process.

and the remaining photoresist is removed (together with the metal covering it). The thin gate is then left between source and drain (Figure 8.9).

In the recessed-channel technique, the source and drain electrodes are defined first and then covered with photoresist. A channel is then etched through the opening in the photoresist until the proper thickness is obtained. The gate is then deposited, and the photoresist and covering metal are removed (Figure 8.10). Selective side etching of the gate metal is used to further reduce the gate length by producing a T-shaped cross section [Smith and Swanson 1989].

Contacts must then be made to the three electrodes. A typical layout is shown in Figure 8.11. To increase the power capability, several identical adjacent cells can be interconnected, either with crossovers or via holes.

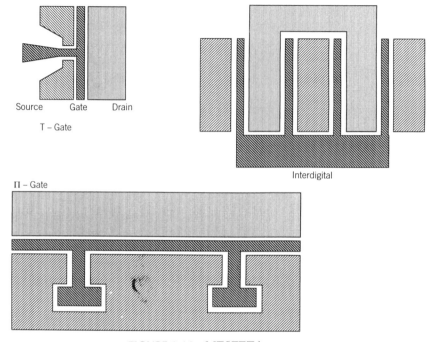

FIGURE 8.11 MESFET layouts.

8.4 AMPLIFIER DESIGN

8.4.1 Basic Purpose

When designing a transistor amplifier, one wishes, first of all, to obtain a specified amplification over a specified frequency band. To obtain it, reflections must be reduced at both transistor ports. The input and output are matched, respectively, to the output of the preceding stage and to the input of the following one. In many situations, matching is done to a normalized value generally taken as 50 Ω, because this is the characteristic impedance of the transmission lines commonly used in microwave applications.

In many amplifiers, one wishes to maximize the gain or the power efficiency. This is obtained at the conjugate match, which occurs when the matching circuit presents the complex conjugate of the transistor's impedance at this port (Section 4.4.7). Note that a transistor is a bidirectional device in which the matching done at the input affects the matching at the output, and vice versa.

When low-noise amplification is required, matching is done to reduce the device noise to its lowest possible value (Section 4.7.7). The input matching circuit must then present the impedance that provides the smallest noise figure (this situation is generally different from both the conjugate and the reflectionless match).

Additionally, the transistor amplifier must be stable; in other words, it must not oscillate at any frequency. This condition must be satisfied with all the loads that might be connected to the amplifier. This requirement is particularly difficult to meet in the design of broad-band amplifiers.

8.4.2 Amplifier Gain

The transducer power gain of the assembly shown in Figure 8.1 is given by [Wroblewski 1988].

$$G_T = \frac{P_L}{P_A} = \frac{\left(1 - |\Gamma_g|^2\right)|\underline{S}_{21}|^2\left(1 - |\Gamma_L|^2\right)}{\left|(1 - \Gamma_g\underline{S}_{11})(1 - \Gamma_L\underline{S}'_{22})\right|^2}, \qquad (8.4)$$

where

$$\underline{S}'_{22} = \underline{S}_{22} + \frac{\underline{S}_{12}\underline{S}_{21}\Gamma_g}{1 - \underline{S}_{11}\Gamma_g}. \qquad (8.5)$$

A convenient approximation is to assume that the transistor is unilateral (i.e., that $\underline{S}_{12} = 0$). Then $\underline{S}'_{22} = \underline{S}_{22}$, and one obtains the unidirectional transducer gain G_{Tu}, which can be maximized by letting $\Gamma_g = \underline{S}^*_{11}$ and $\Gamma_L = \underline{S}^*_{22}$:

$$G_{umax} = \frac{P_L}{P_A} = \frac{|\underline{S}_{21}|^2}{\left(1 - |\underline{S}_{11}|^2\right)\left(1 - |\underline{S}_{22}|^2\right)}. \qquad (8.6)$$

The maximum gain is obtained with matching circuits presenting, respectively, $\Gamma_g = \underline{S}^*_{11}$ at the input and $\Gamma_L = \underline{S}^*_{22}$ at the output. The circuits are realized with sections of transmission lines, which may have different characteristic impedances, and open-ended stubs (reactive matching, Section 5.2.1). In general, an approximate circuit design is first made on the Smith chart (Section 4.6.6) at the center frequency and used as starting point for a computer-aided design (CAD) optimization process (Section 12.2.4). The design of broad-band amplifiers requires more complex circuits, with several matching sections [Matthaei et al. 1964; Abrie 1986; Quirarte 1991]. More elaborate broad-banding schemes include resistors, feedback, and active matching [Wroblewski 1988]. Distributed amplifiers, in which amplifiers are inserted as coupling elements between two transmission lines, provide much wider operating bandwidths than are achievable with single amplifiers

[Cooper et al. 1989]. Finally, it is also possible to combine several amplifier sections to increase the gain and the power output capability.

Suitable dc biasing must be provided to the transistors by the input and output circuits. Capacitors are inserted so that no dc can flow into the remainder of the microstrip circuit, while precautions are required to prevent the microwave signal from reaching the dc bias supplies.

8.4.3 Stability

An amplifier should not oscillate, no matter what load is connected to it. This means that the real parts of the input and output impedances should not become negative. These conditions define the *stability circle* on the Smith chart [Parisot and Soares 1988]. On the load side, the center and radius of the circle are given by

$$\underline{C}_{stL} = \frac{\underline{S}_{11}\underline{\Delta}^* - \underline{S}_{22}^*}{|\underline{\Delta}|^2 - |\underline{S}_{22}|^2}, \qquad R_{stL} = \frac{|\underline{S}_{12}\underline{S}_{21}|}{|\underline{\Delta}|^2 - |\underline{S}_{22}|^2}, \qquad (8.7)$$

where

$$\underline{\Delta} = \underline{S}_{11}\underline{S}_{22} - \underline{S}_{12}\underline{S}_{21}. \qquad (8.8)$$

To determine whether the stable region is located inside or outside of the stability circle, one determines whether the operation of the transistor is stable when it is terminated into a matched load. For a unilateral transistor, $\underline{S}_{12} = 0$ and the circle shrinks to a point.

The stability circle at the input is defined in a similar manner, yielding a stability circle whose center and radius are given by

$$\underline{C}_{stg} = \frac{\underline{S}_{22}\underline{\Delta}^* - \underline{S}_{11}^*}{|\underline{\Delta}|^2 - |\underline{S}_{11}|^2}, \qquad R_{stg} = \frac{|\underline{S}_{12}\underline{S}_{21}|}{|\underline{\Delta}|^2 - |\underline{S}_{11}|^2}. \qquad (8.9)$$

Under particularly favorable conditions, the two circles cover the complete Smith chart, and the transistor is said to be *unconditionally stable*. No spurious oscillations can occur when the transistor's scattering parameters satisfy the following conditions:

$$K = \frac{1 - |\underline{S}_{11}|^2 - |\underline{S}_{22}|^2 + |\underline{\Delta}|^2}{2|\underline{S}_{12}\underline{S}_{21}|^2} > 1,$$

$$|\underline{\Delta}| = |\underline{S}_{11}\underline{S}_{22} - \underline{S}_{12}\underline{S}_{21}| < 1. \qquad (8.10)$$

Instability is directly related to the presence of internal feedback. Stability conditions must be determined at several frequencies within the operating band of the amplifier, but also outside of the band, to make sure that the transistor cannot oscillate at all. These requirements may be difficult to meet in broad-band amplifier design [Jung and Wu 1990].

8.4.4 Noise

An amplifier not only amplifies the input signal, it also contributes some additional noise. The result is that the signal-to-noise ratio decreases as a signal is amplified: any signal amplification produces some degradation.

Since the noise is often of thermal origin, its power is expressed in terms of an equivalent blackbody temperature (a blackbody absorbs all incident radiation and reemits it entirely and evenly distributed over the frequency spectrum)

$$N = k_B TB, \qquad (8.11)$$

where N is the average noise power in watts, $k_B = 1.3804 \times 10^{-23}$ J/K is the Boltzmann constant, T is the equivalent noise temperature in kelvins, and B is the frequency bandwidth in hertz. This expression is only valid when $k_B T \gg hf$, where $h = 6.62 \times 10^{-34}$ J-s is Planck's constant and f is the frequency. The equivalent noise figure of an amplifier is determined by measuring the noise power added by the amplifier.

The noise figure F is defined by the ratio of the signal-to-noise ratios at the output and input of the device (amplifier) when the input is connected to a noise source at the standard noise temperature $T_0 = 290$ K.

$$F = \frac{(S/N)_{\text{input}}}{(S/N)_{\text{output}}} = 1 + \frac{T}{T_0}. \qquad (8.12)$$

It is often expressed in decibels:

$$LF = 10 \log_{10} F. \qquad (8.13)$$

An ideal lossless amplifier would not add any noise power. Its noise temperature would be $T = 0$ K, providing a noise figure $F = 1$ or $LF = 0$. The lower the noise figure, the better the amplifier.

The noise figure of a transistor depends directly on the admittance of the source connected at the transistor's input. Its smallest possible value F_{min} is obtained for a particular value of the admittance $\underline{Y}_{\text{min}}$. It varies as a function of the source admittance $\underline{Y}_s = G_s + jB_s$:

$$F = F_{\text{min}} + \frac{R_n}{G_s} |\underline{Y}_s - \underline{Y}_{\text{min}}|^2, \qquad (8.14)$$

where R_n is the noise resistance, which indicates the rate of increase of the noise figure since the source is not optimally matched for low-noise operation. The points corresponding to constant values of the noise figure form circles on the Smith chart.

Several mechanisms produce noise. Thermal noise is contributed by all lossy elements. A noise component proportional to $1/f$ is particularly bothersome at low frequencies. Shot noise results from the granular nature of the current flow that crosses a junction. Since a bipolar transistor possesses two junctions through which the entire current flows, while MESFETs and HEMTs only have one negatively biased gate junction (small reverse current), the latter possess a significant inherent advantage over the bipolar transistor. HEMTs exhibit particularly small noise figures that can be reduced by cooling, and they can be used to realize receivers with noise performances similar to those of parametric amplifiers.

Low-noise amplification is required in all applications where signals of very small amplitude must be amplified, conditions most often encountered in the input stage of receivers for satellite and space communications.

8.4.5 Efficiencies

The efficiency of an amplifier indicates which part of the power fed to it has actually come out as useful output power. In conventional electronics, efficiency is simply defined as the ratio of the output power to the power supplied by the biasing sources. Because the gain is large, the input power is quite small and not considered. At microwaves, the gain can become rather low, down to 10 dB or less, and efficiency can then be defined in three ways:

1. The total efficiency is the ratio of the output power to all the power supplied to the amplifier (input + bias):

$$\eta_t = P_{out}/(P_{in} + P_b), \qquad (8.15)$$

where P_{out} is the output power, P_{in} the input power, and P_b the power supplied by the bias sources.

2. The partial efficiency does not consider the power of the input signal:

$$\eta_p = P_{out}/P_b. \qquad (8.16)$$

3. The added power efficiency is given by the difference between output and input power divided by the bias power:

$$\eta_{pa} = (P_{out} - P_{in})/P_b. \qquad (8.17)$$

It can readily be seen that

$$\eta_p \geq \eta_t \geq \eta_{pa}. \tag{8.18}$$

The three quantities are equal when the amplifier gain is infinite. The difference between the three quantities increases as the gain decreases.

8.4.6 Active Circulators

Combining amplifiers with sections of transmission lines or reactive elements makes it possible to build up active circulators [Ayasli 1989; Hara et al. 1990; Katzin et al. 1992]. The resulting devices that can be built up in MMIC technology are smaller, lighter, and able to operate over wider bandwidths than the corresponding ferrite circulators (Section 6.3.3). They are, however, active devices, which require a dc bias for their operation.

8.5 OSCILLATOR DESIGN

8.5.1 General Principle

Basically, an oscillator is made with an amplifier in which a part of the output signal is fed back to the input, with suitable amplitude and phase conditions. The feedback circuit must be strongly frequency selective to ensure that at any given time the oscillation can only occur at a single frequency.

A patch (Section 7.1) or a dielectric resonator (Section 7.2) can be used in the feedback circuit, in which case the frequency of the output signal remains constant (fixed-frequency oscillator). More stable oscillating characteristics are provided by dielectric resonators than by microstrip patches, since the latter have a much lower quality factor.

Frequency-tuned signal sources can be realized with frequency-dependent resonators. The resonant frequency of a YIG (yttrium–iron–garnet, Section 13.1.10) sphere can be magnetically swept over a very wide frequency range. However, the equipment required for the magnetic bias can be bulky, and the response is rather slow. A varactor adjustable resonator can also be used (Section 7.1.6), in which the capacitance is voltage dependent. It provides a faster response and is more compatible with microstrip structures, but the tuning range is narrower.

8.5.2 Negative Input Resistance

The internal feedback of the transistor can be used (series feedback). As pointed out in Section 8.4.3, some transistors are unstable for certain generator impedance conditions, in which case the transistor exhibits a

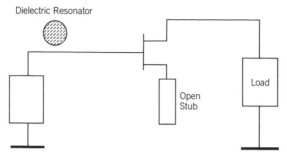

FIGURE 8.12   Series-feedback transistor oscillator.

negative resistance. This situation is highly undesirable for amplifiers, but can be put to good use in the design of oscillators. The negative input resistance can further be enhanced by connecting a pure reactance (open line) to the source of the transistor (Figure 8.12). In this geometry the resonator is coupled to the input line. The spacing is determined to provide a suitable coupling, and the distance from the transistor provides the correct phase shift for oscillation.

8.5.3   External Feedback

An alternative approach provides external coupling from the transistor's output to its input, across the resonant circuit (Figure 8.13). In this case also, suitable coupling must be provided between the two lines and the resonator to ensure that the loop gain is larger than unity. The lengths of the lines must also be carefully adjusted to provide feedback with the proper phase so that the complete phase shift around the loop is a multiple of 2π.

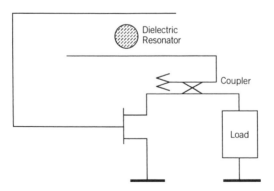

FIGURE 8.13   Parallel-feedback transistor oscillator.

8.5.4 Comparison

The series-feedback approach provides basically a simpler geometry, because only one line couples to the resonator. The adjustment is thus inherently simpler. The parallel-feedback structure inherently takes more space, because some lengths of microstrip line are required to provide proper phases. The resonator is coupled on both sides, so couplings are more difficult to adjust. When the amplifier gain is large, the resonator coupling may be significantly reduced, resulting in a higher loaded Q factor and a better frequency stability.

8.6 PROBLEMS

8.6.1 Determine the power gain at 10 GHz of a transistor, knowing that its input impedance is $10 + j12\ \Omega$, its output impedance is $51 - j23\ \Omega$, and its cutoff frequency is 12 GHz.

8.6.2 What is the cutoff frequency of a MESFET having a gate length of $0.25\ \mu\text{m}$ and made in a material with a saturation velocity of 1.3×10^7 cm/s?

8.6.3 What is the noise power of a device at a temperature of 40 K having a bandwidth of 200 MHz?

8.6.4 Calculate the three efficiencies for a transistor, knowing that its power gain is 8, that it is fed an input power of 1 W, and that the biasing circuits supply altogether 18 W.

CHAPTER NINE

Solid-State Devices

Signal processing can be carried out by solid-state devices inserted in a microstrip circuit. Diodes can detect a signal, attenuate it, switch it, shift its phase, modulate it, or still mix it with other signals. The frequency can also be shifted or multiplied. Furthermore, a signal can be generated directly within the microstrip circuit by two-terminal oscillators, making use either of transferred electron devices or of avalanche effects. Diodes are used as nonlinear components, as voltage-controlled impedances, and as microwave sources [Chang 1990; Yngvesson 1991].

9.1 DIODES

9.1.1 General Description

Two-terminal semiconductor devices, called diodes, contain a rectifying junction either between a metal and a semiconductor (Schottky barrier diodes, Section 9.1.2) or between two differently doped regions of the semiconductor material (*pn* diode, Section 9.1.3). In microwave diodes, free charge carriers (electrons or holes) cross a junction in a time shorter than the period of the signal. The area of the junction must also be small enough to keep the parasitic capacitance at a low value.

9.1.2 Schottky Barrier Diode

In Schottky barrier diodes, also called hot-carrier diodes, a signal is rectified or mixed across a metal–semiconductor interface (Figure 9.1). The semiconductor may be silicon (Si) or gallium arsenide (GaAs).

Schottky diodes operate on the same principle as the old point-contact detectors, in which a metallic cat whisker contacts a semiconducting crystal. A much better control of the characteristics is, however, attained with

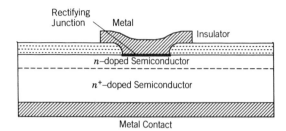

FIGURE 9.1   Geometry of a Schottky barrier diode.

present-day planar manufacturing techniques. Charges build up on both sides of the metal–semiconductor interface, forming a dipole layer which creates a potential barrier about $\frac{1}{2}$ to 1 V high.

9.1.3   *pn* Junctions

Within a semiconductor material (Si, GaAs), a *pn* junction separates a region doped with acceptor impurities (*p*-region) from a region doped with donor impurities (*n*-region). Starting with a slice of *n*-doped material, one dopes an area with *p*-type impurities by diffusion or ion implantation [Sze 1981]. Metallic ohmic contacts are then deposited on both sides (Figure 9.2).

9.1.4   *I-V* Characteristic

The current crossing a junction has a nonlinear dependence on the voltage across the junction. For a Schottky barrier junction and for a *pn* junction between lightly doped semiconductors, the general behavior is sketched in Figure 9.3.

Current flows when the junction is biased in the forward direction. When it is reverse biased, only a trickle of current can get through the junction,

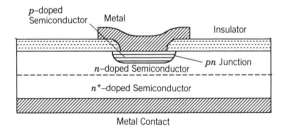

FIGURE 9.2   Geometry of a *pn* diode.

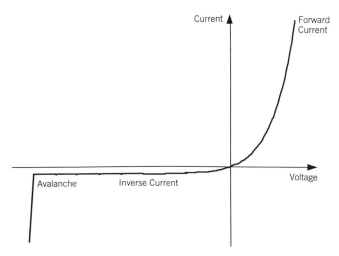

FIGURE 9.3   *I-V* characteristic of a rectifier diode.

most usually negligible until the reverse breakdown voltage is reached. The reverse current then increases rapidly, and precautions must be taken to keep it at a reasonable level. Avalanche breakdown can be used to generate microwave signals (Section 9.5.3).

For low signal amplitudes, the nonlinear current-to-voltage characteristic takes the general form [Sze 1981]

$$I = I_s \left[\exp\left(\frac{qU}{\eta kT} \right) - 1 \right] = I_s [\exp(\alpha U) - 1]$$

$$= I_s \left[\alpha U + \frac{(\alpha U)^2}{2} + \frac{(\alpha U)^3}{3!} + \frac{(\alpha U)^4}{4!} + \cdots \right], \qquad (9.1)$$

where q = charge of the electron
$\quad k$ = Boltzmann constant
$\quad U$ = voltage across the junction
$\quad T$ = physical junction temperature (in kelvins)
$\quad \eta$ = ideality factor (close to 1 for a "good" diode at room temperature
$\quad I_s$ = saturated inverse current
$\quad \alpha = q/\eta kT$

While the *I-V* characteristics of Schottky barrier diodes and of *pn* junctions are similar, the carrier-transfer mechanisms are basically different for the two devices. They rely, respectively, on majority and minority carriers; hence, Schottky barrier diodes have a faster response than *pn* diodes.

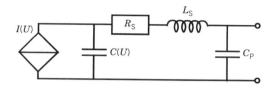

FIGURE 9.4   Equivalent circuit of a diode detector.

9.1.5   Equivalent Circuit

The current $I(U)$ flowing across the junction can be represented by a current source controlled by the voltage. In addition, the junction contributes a shunt nonlinear capacitance $C(U)$. The semiconductor substrate provides a series resistance R_s, while the connecting wires add an inductance L_s. Finally, the ceramic package of an encapsulated diode is represented by an additional shunt capacitance C_p (Figure 9.4).

The equivalent circuit is actually a low-pass filter cell (Section 7.3).

The junction capacitance $C(U)$ results from the presence of fixed charges within the depletion layer [Howes and Morgan 1978], whose width increases with the inverse biasing voltage. The capacitance therefore varies with voltage as

$$C(U) = \frac{C_0}{(1 - U/\Phi)^{\gamma}} \qquad \text{for } U < \Phi, \tag{9.2}$$

where C_0 is the capacitance at zero bias and Φ is the barrier potential. The exponent γ depends on the doping profile of the junction. It is close to 0.5 for Schottky barrier diodes and abrupt pn junctions, while decreasing to $\frac{1}{3}$ for linearly doped pn junctions.

To reduce the substrate resistance as much as possible, a heavily doped n^{++} semiconductor substrate is used as basis material on which a lightly doped n layer is grown by epitaxy (Figures 9.1 and 9.2) [Brodie and Muray 1992].

The wiring inductance can be reduced and the package capacitance suppressed by directly bonding unencapsulated chips on an MIC substrate (Section 9.1.10).

9.1.6   Varactors

When a diode is reverse biased, the inverse current I_s that flows across it is practically negligible, and the device behaves as a variable capacitor with a capacitance given by Eq. 9.2. Diodes specifically designed for use as capacitors, called varactors, are not meant to carry conduction current. Their voltage range extends from the reverse breakdown (avalanche, lowest capacitance value) to a slight forward bias ($U < \Phi$, largest capacitance value).

The series resistance of the substrate determines the cutoff frequency of the varactor at a given voltage

$$f_c(U) = \frac{1}{2\pi R_s C(U)}.$$ (9.3)

The quality factor is defined at a specified biasing voltage and operating frequency f_0 by

$$Q(U) = f_c(U)/f_0.$$ (9.4)

9.1.7 Tunnel and Backward Diodes

When the semiconductor on both sides of a *pn* junction is heavily doped, the junction becomes very narrow and carriers can "tunnel" through the junction. The resulting *I-V* characteristics are shown on Figure 9.5. The characteristic of a tunnel diode is shaped like the letter *N*, with a negative resistance in the center part. A backward diode behaves like a regular *pn* diode in the forward part of the characteristic, while the current increases more rapidly with reverse bias.

9.1.8 *pin* Diodes

Considering the forward *I-V* characteristic of a diode (Figure 9.3), it is apparent that it can also be used as a voltage-controlled resistor. Its internal resistance is large for a reverse bias and decreases when the diode is forward

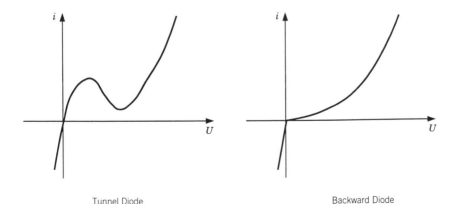

Tunnel Diode   Backward Diode

FIGURE 9.5 *I-V* characteristic of a tunnel and a backward diode.

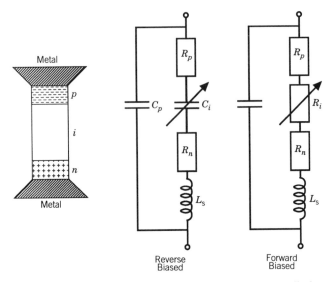

FIGURE 9.6 Structure and equivalent circuit of a *pin* diode.

biased. A diode can thus be used to realize variable attenuators, amplitude modulators, and switches (Section 9.4).

The variable-resistor characteristic can be enhanced by increasing the width of the depletion layer, which is done by adding a layer of intrinsic (i.e., undoped) semiconductor, producing in this manner a *pin* diode (Figure 9.6). In practice, it is quite difficult to obtain really intrinsic semiconductor materials, so the *i*-region is more likely to be a very lightly doped *n*- or *p*-region [White 1990].

An inversely biased *pin* diode is almost a capacitor: the intrinsic region is the dielectric, the more heavily doped *p*- and *n*-regions the electrodes. With forward bias, the intrinsic region becomes conductive since carriers are injected into it from both sides. The *pin* diode is a current-controlled resistor.

9.1.9 IMPATT Diodes

When the inverse voltage across a *pn* junction is large, minority carriers in the junction receive enough energy from the field to ionize surrounding atoms by impact. The process builds up like a chain reaction, or avalanche, and the current increases rapidly with time (Figure 9.3). It must be limited by the surrounding circuit, or the process might destroy the diode.

When the breakdown, or avalanche, voltage is applied to the junction, the current resulting from the impact ionization process increases exponentially with time. There is, therefore, a slight delay between voltage and current, as

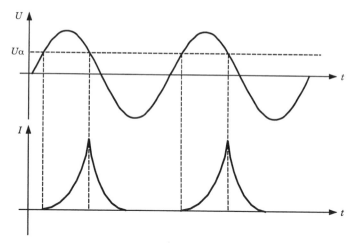

FIGURE 9.7   Delay between current and voltage in the avalanche region.

shown in Figure 9.7 for a small alternating current superimposed on a negative dc bias.

The carriers produced in the junction are then made to cross a charge-free drift zone, producing additional delay between U and I, until the current flowing in the circuit is 180° out of phase with respect to the voltage. One obtains in this manner a negative resistance, and the diode can be used to realize a microwave generator (Section 9.5.3) [Gibbons 1973].

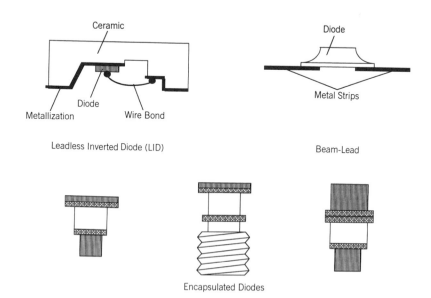

FIGURE 9.8   Diode packages available. (Courtesy High Tech Tournesol.)

9.1.10 Diode Packaging

Diodes of most types are available for mounting on microstrip. Diode chips can be directly inserted or encased in various packages: leadless inverted device (LID), beam lead (connection with thin strips) or cylindrical cases, with or without mounting screw (Figure 9.8). To mount the last package, a hole must be drilled through the substrate (Section 13.6.7) so that the lower part of the diode is in close contact with the ground plane, which ensures heat removal.

9.2 DETECTORS

9.2.1 Rectifiers

Every microwave or millimeter-wave system eventually ends in some kind of detector, which rectifies the high-frequency signal. A dc or a low-frequency signal is then obtained, with an amplitude corresponding to the envelope of the high-frequency signal. The information carried by a modulated signal can thus be retrieved and processed [Kollberg 1990].

Rectifiers used at microwave and millimeter-wave frequencies are generally Schottky barrier diodes (Section 9.1.2). *pn* diode rectifiers (Section 9.1.3) have a slower response and are used over the lower frequency ranges. Tunnel diodes and backward diodes (Section 9.1.7) are used as sensitive low-noise detectors over the microwave range.

9.2.2 Rectifier Operation: Square-Law Detector

When a sine-wave voltage $U = U_1 \sin \omega t$ is applied across a junction, the current flowing through the junction varies periodically, but its shape is not a sine wave (Figure 9.9).

The current has an asymmetrical behavior, which denotes the presence of a dc component I_0 contributed by the even components in Eq. 9.1:

$$I_0 = I_s \left[\frac{(\alpha U_1)^2}{4} + \frac{(\alpha U_1)^4}{64} + \frac{\alpha U_1^6}{2304} + \cdots \right]. \tag{9.5}$$

Under normal operating conditions the quadratic term (in U_1^2) is the only significant one: the dc component of the current is proportional to the square of the microwave voltage across the diode's junction and, thus, to the signal's power. The rectifier is then called a *square-law detector* and is characterized by a microamp per milliwatt conversion factor. Measuring instruments are calibrated in terms of the quadratic behavior.

In addition, the current flowing across the junction contains time-varying components at all the harmonic frequencies $n\omega$. The detector diode is meant to extract only the dc component and is therefore followed by a low-pass

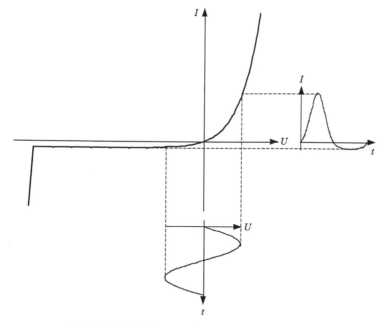

FIGURE 9.9 Voltage and current across a rectifier.

filter (Section 7.3) which blocks the time-dependent components of the current.

9.2.3 Noise and Tangential Sensitivity

The detector operation is limited at low signal levels by various kinds of noise produced by the diode. Shot noise results from the granular nature of the current flow within the diode. A noise component proportional to $1/f$ is particularly important at low frequencies. Thermal noise is contributed by the substrate resistance, while hot-electron noise becomes significant in Schottky diodes with a large forward bias. Shot noise and thermal noise can be reduced by cooling the detector.

The tangential sensitivity of a detector is defined as the input power required to increase the detector's output by an amount equal to the voltage noise fluctuations. It is generally determined by looking at the detector's output on an oscilloscope.

9.2.4 Dynamic Range

Detectors can be used only over a limited range of input powers. The presence of noise (Section 9.2.3) sets a lower bound. A signal at a power level below the average noise power cannot be detected accurately with a simple diode.

At larger power levels the *I-V* characteristic of the diode is no longer quadratic, so Eq. 9.1 is no longer valid. The current becomes proportional to the voltage (i.e., to the square root of the power for the next range of input powers) and it then saturates. Power measurements outside of the quadratic region require a calibration of the measuring equipment over the complete power range. It is therefore preferred to keep the signal within the quadratic region of the detector, if necessary by lowering its level with known attenuators.

The application of an excessive power level to a detector diode may physically destroy it, and protections are needed when a sensitive receiver is close to a transmitter, as for example, in a monostatic radar [Gardiol 1984].

9.3   MIXERS

9.3.1   Frequency Mixing

In a mixer, two time-dependent signals are simultaneously applied to a semiconductor diode (Schottky barrier diode, Section 9.1.2, or *pn* diode, Section 9.1.3) characterized by a nonlinear *I-V* characteristic (Eq. 9.1). We consider two sine waves with angular frequencies ω_1 and ω_2:

$$U = U_1 \sin \omega_1 t + U_2 \sin \omega_2 t. \qquad (9.6)$$

The current I flowing through the diode then contains components having all the combinations of angular frequencies $|m\omega_1 + n\omega_2|$, with $m, n = 0, 1, 2 \ldots$, which are called the intermodulation products. The main contribution comes from the quadratic term in Eq. 9.1, which is developed as follows:

$$I_s \frac{(\alpha U)^2}{2} = I_s \frac{\alpha^2}{2} \left(U_1^2 \sin^2 \omega_1 t + U_2^2 \sin^2 \omega_2 t + 2U_1 U_2 \sin \omega_1 t \sin \omega_2 t \right)$$

$$= I_s \frac{\alpha^2}{4} \left[U_1^2 + U_2^2 - U_1^2 \cos 2\omega_1 t - U_2^2 \cos 2\omega_2 t \right.$$

$$\left. + 2U_1 U_2 \cos(\omega_1 - \omega_2)t - 2U_1 U_2 \cos(\omega_1 + \omega_2)t \right]. \qquad (9.7)$$

One component is at the difference of the two angular frequencies, $\Delta\omega = |\omega_1 - \omega_2|$. Since the cosine is an even function, the same value of $\Delta\omega$ is obtained in two situations: when $\omega_2 = \omega_1 + \Delta\omega$ and when $\omega_2 = \omega_1 - \Delta\omega$ (Section 9.3.3).

9.3.2   Mixer Operation

In a mixer, a small-amplitude signal at $\omega_1 = \omega_s$ is mixed with a larger signal at $\omega_2 = \omega_{lo}$ provided by a local oscillator (lo, sometimes called a pump signal). An intermediate-frequency (if) signal appears at the mixer's output:

$$\Delta\omega = \omega_{if} = |\omega_s - \omega_{lo}| = |\omega_1 - \omega_2|. \tag{9.8}$$

The mixer yields a frequency-shifted signal of average amplitude. It "slides" signals from one frequency band to another, generally downward in frequency. Signals are brought in this manner into frequency ranges where signal processing is done with standard electronic equipment [Howes and Morgan 1978; Kollberg 1984; Maas 1986].

The signals at the three frequencies must be clearly separated to avoid possible interference in other nonlinear devices: the three ports of the mixer are therefore connected to band-pass filters, which are tuned at ω_s, ω_{lo}, and ω_{if}. Complex interactions involving the different intermodulation products take place within the mixer. The filters must therefore be adjusted to provide adequate reflections at the corresponding frequencies. A similar situation is encountered in frequency multipliers (Section 9.6).

9.3.3 Image Frequency

As indicated in Section 9.3.1, a receiver with a mixer connected to a local oscillator at ω_{lo} and with an output filter tuned at ω_{if} can receive signals at frequencies $\omega_s = \omega_{lo} + \omega_{if}$ and $\omega_s = \omega_{lo} - \omega_{if}$. Obviously, a receiver should be tuned to a single frequency only, so one of the two inputs is unwanted. The frequency of the latter is called image frequency, and the two possible situations are shown in Figure 9.10.

In Figure 9.10a, the signal frequency is smaller than that of the local oscillator (lower sideband), while the situation is reversed in Figure 9.10b

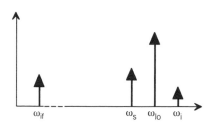

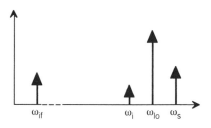

FIGURE 9.10   Relative positions of the frequencies in a mixer: (a) lower sideband, (b) upper sideband.

(upper sideband). For proper receiver operation, the image frequency should be carefully suppressed, for instance, by locating a pole of attenuation of the band-pass filter precisely at that frequency. One could also use a band-stop filter.

9.3.4   Balanced Mixer

The signal and the local oscillator signal must be fed together to the mixer diode but kept adequately isolated from each other to prevent unwanted feedback. This can be done by using a branch line coupler (Section 6.5.1) or a hybrid ring (Section 6.5.3) with two mixer diodes. In this manner one obtains a balanced mixer (Figure 9.11).

Both assemblies provide separate input ports for the signal and for the local oscillator, with some isolation between the two. The if signals across the two diodes are in quadrature when using a branch line coupler, in opposition with a hybrid ring, and must be combined accordingly at the input of the if amplifier. In both cases, the lo noise at the if frequency tends to cancel out, thereby increasing receiver sensitivity. The isolation and the amount of noise suppression depend on the actual power balance of the coupler, while the diodes should be as close to identical as possible. The branch line coupler assembly provides a better input match (as diode mismatches tend to cancel out), while the hybrid ring assembly yields a higher isolation.

9.3.5   FET Mixer

While common mixer designs make use of semiconductor diodes, mixing can also be done with transistors, thus combining mixing and amplification within the same device. Since some of the equivalent-circuit elements of the FET are bias dependent (Section 8.3.4), when a low-level signal is applied to a FET pumped by a strong lo signal, mixing occurs and intermodulation products are produced at the combinations of angular frequencies $|m\omega_1 + n\omega_2|$.

Several schemes can be used: either both the signal and the local oscillator are connected to the FET gate, or the signal is connected to the gate and the

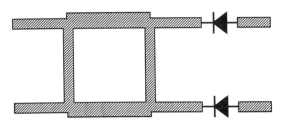

FIGURE 9.11   Microstrip mixer.

local oscillator to the drain. Dual-gate FETs can also be used, or two single-gate FETs can be series-connected [Kollberg 1984].

9.3.6   Harmonic Mixer

Mixers can also operate with higher-order intermodulation products, generally by mixing the signal with a harmonic multiple of the local oscillator (pump) signal. The intermediate frequency of a harmonic mixer is then defined by

$$\omega_{if} = |\omega_s - n\omega_{lo}| \qquad \text{with } n = 2, 3, \dots . \tag{9.9}$$

The band-pass filters at the three ports must be tuned accordingly.

9.4   CONTROL DEVICES

9.4.1   *pin* Diode Attenuator

A forward-biased *pin* diode is, basically, a variable resistor controlled by a bias current (Section 9.1.8). It can be inserted into a microstrip circuit to realize an attenuator. However, since its impedance is not constant, matching of the resulting device is difficult. It is possible to use two identical elements, separated by a quarter-wave transmission line, so that the reflections of the two devices cancel each other.

When used as a variable resistor, the *pin* diode may dissipate a significant part of the input power, and the attenuator must be designed accordingly.

9.4.2   *pin* Diode Switch

The *pin* diode resistance varies from a very low value ($\cong$ short circuit) to a very large value ($\cong$ open circuit). At both extremes, the diode reflects practically all the incident signal and thus absorbs little power. By connecting

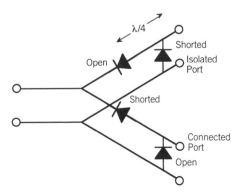

FIGURE 9.12   Schematic of a *pin* diode switch.

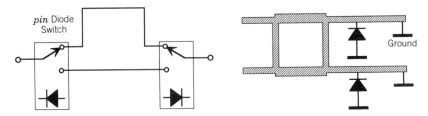

FIGURE 9.13   *pin* diode phase shifters.

pin diodes on a microstrip line in shunt or in series, one obtains a switch (Figure 9.12). Practical designs take into account the parasitic elements of the diode in the matching process.

9.4.3 *pin* Diode Phase Shifter

pin diode switches combined with sections of transmission line having different lengths are used to realize digital phase shifters (Figure 9.13).

9.4.4 Varactor Phase Shifter

A similar approach with variable reactive elements (varactors) is used to realize continuously variable phase shifters. A transmission line terminated on a varactor is an almost complete reflection, the phase of which depends on the bias. Combining two identical varactors with a branch line coupler provides an analog (continuously variable) phase shifter [Sreenivas and Stockton 1990].

9.5 SIGNAL SOURCES

9.5.1 The Gunn Effect

Experiments with III-V semiconductor materials led to the discovery of current instabilities in GaAs and InP [Gunn 1964]. The observed phenomenon is explained by the existence of two "valleys" in the energy-versus-momentum diagram of the materials. The low-energy electrons located in the lower valley possess high mobility. More energetic ones in the upper valley move more slowly. When the electric field in the semiconductor exceeds a specific threshold level, electrons moving through the semiconductor jump from the lower valley into the upper one, and their speed is reduced accordingly (electron transfer). The resulting situation is similar to the one observed when cars slow down on a highway. Electrons group into a packet, called a domain, that moves across the semiconductor with the upper valley speed (slower). The electric field concentrates within the domain, and in most devices only one domain can exist at any given time.

No current flows in the outside circuit while a domain is formed and travels across the device. When it reaches the far side of the semiconductor, a pulse of current flows into the circuit, while another domain starts to form. The process is repeated periodically, with a repetition rate that is inversely proportional to the thickness of the material.

A proper selection of material doping and thickness yields a simple bar of semiconductor which, when biased with a dc voltage, becomes a microwave source. This is in considerable contrast with previously available microwave sources, which are tubes that require complex wiring and several highly stabilized voltage supplies (klystrons, backward-wave oscillators (BWOs).

9.5.2 Transferred Electron Oscillators

The efficiency of the simple semiconductor source described in Section 9.5.1 is rather low, but it can be increased by combining the source with a resonator, which also provides frequency stabilization and tuning. Several modes of operation can be achieved, depending on the carrier concentration, semiconductor length, resonator frequency, and coupling: resonant, delayed, quenched, limited space-charge accumulation (LSA), hybrid modes [Carroll 1970]. A transferred electron source with a microstrip patch resonator (Section 7.1) is shown in Figure 9.14.

9.5.3 IMPATT Diode Oscillators

A *pn* diode biased in the avalanche range with a drift region to provide a carrier delay provides a negative resistance (Section 9.1.9). Introducing an IMPATT (impact avalanche transit-time) diode within a resonant structure

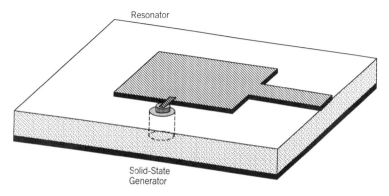

FIGURE 9.14 Semiconductor source on microstrip.

(Figure 9.14) also provides a simple way to realize a microwave signal generator [Gibbons 1973; Howes and Morgan 1976].

9.5.4   Frequency Tuning

Microwave signal generators can be tuned over a limited frequency range by electronic means by incorporating a varactor (variable capacitance, Section 9.1.6) into the resonant circuit. Separate biasing voltages must be provided for the signal generator and for the varactor [Howes and Morgan 1978].

9.5.5   Comparison

Transferred electron effect devices (TEDs) and IMPATT diodes both provide simple means to generate microwave signals within a microstrip circuit. However, their impedance levels and required biasing voltages are different. TEDs typically produce powers of a few milliwatts, whereas IMPATT diodes yield tens of milliwatts. The IMPATT signal is noisier than the TED signal [Yngvesson 1991]. Both devices are being increasingly replaced by transistors within the microwave range (Chapter 8), although they are mostly used over the millimeter-wave range.

9.5.6   Frequency Multipliers

Microwave and millimeter-wave signals can also be generated by frequency multiplication of lower-frequency signals, for instance of highly stable signals produced by a quartz-controlled generator. A signal of angular frequency ω_0 is fed to a strongly nonlinear device, which generates harmonics at angular frequencies $n\omega_0$. The output signal at the desired frequency is then selected by a band-pass filter (Figure 9.15).

Another band-pass filter is connected to the input port to keep the harmonics out of the low-frequency circuit.

The nonlinear device providing the harmonic generation can be resistive or reactive. A resistive element uses the nonlinear I-V characteristic of a diode (Eq. 9.1). It absorbs part of the input signal, so the multiplier efficiency cannot exceed $1/n^2$ [Page 1956]. On the other hand, the theoretical efficiency of an ideal reactive (lossless) multiplier is 100% [Manley and Rowe 1956].

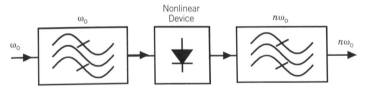

FIGURE 9.15   Frequency multiplier.

However, this is never achieved in practice due to losses in the nonlinear element (varactor, Section 9.1.6), filters, and connecting circuit.

The interactions that take place within the nonlinear device are quite complex, involving not only the frequency multiplication of the input signal but also multiple mixing with harmonic components at $k\omega_0$ that are reflected by the filters on both sides. For proper operation of the multiplier, the filters must not only be matched over their respective passbands, but they must also provide purely reactive terminations (completely reflecting) with adequate phases for the other harmonic components [Uenohara and Gewartowski 1969]. A similar situation is encountered in mixers (Section 9.3).

Frequency multiplication by a low value of n is generally accomplished with varactors (Section 9.1.6). For larger multiplication factors one uses a step diode (pulse recovery, snap diode), which provides a very short and sharp pulse when switched from forward- to reverse-bias conditions [Friis 1967]. Multiplication can also be obtained with transistors (class B or C amplifiers).

9.6 PROBLEMS

9.6.1 A varactor has an inverse voltage limit of -50 V and can be forward-biased up to $\phi/2$ without significant increase in current ($\phi = 0.7$ V). Determine the capacitance tuning ratio (max/min) that can be achieved when $\gamma = 0.5$, 0.35, and 0.25.

9.6.2 For which value of the voltage U_1 in a rectifier will the fourth-power term become significant (5% of the quadratic term)?

CHAPTER TEN

Mathematical Techniques

Electromagnetic analysis requires the resolution of Maxwell's equations in the presence of the boundary conditions at interfaces. Since microstrip structures are three times inhomogeneous (Section 1.5.2), this leads to very complex developments, even for rather simple geometries. Over the years, a wide variety of techniques has been proposed for their analysis, going from crude expressions established by considering simplified models, all the way to sophisticated integral equation formulations involving Green's functions, which can now be efficiently treated with numerical techniques on powerful supercomputers.

10.1 PRINCIPLES FOR A CLASSIFICATION

10.1.1 Introduction

Microstrip structures are many and varied, including simple transmission lines, rectangular and circular resonators, and patch antennas, and more complex structures such as discontinuities, junctions, couplers, and filters. Many different approaches, at various levels of approximation and sophistication, were developed to analyze the large variety of microstrip structures encountered in applications [Itoh 1989; Sorrentino 1989; Yamashita 1990].

First of all, simple criteria for classification are required to allow one to determine in which category the problem to be solved is located and to select the approaches most likely to succeed in its resolution. The principles underlying the main mathematical techniques used to describe and analyze microstrip structures are then briefly outlined.

Classification is particularly difficult because some approaches that have different names are closely related, while others have similar names but are very different. The following sections try to present an overall view of the techniques most often encountered in the analysis of microstrip.

10.1.2 Quasi-TEM or Full-Wave?

Obviously, the determination of the dc behavior of a microstrip structure presents little interest in itself, because all microstrips are intended for the transmission of signals that inherently present a time-dependent behavior. Still, the elements governing the operation of microstrip over the low-frequency ranges are the capacitances and inductances that make its equivalent circuit in the *quasi-TEM* approximation (Section 4.5), and these are best determined by means of static and quasi-static approximations.

In the static case, the time derivatives in Maxwell's curl equations vanish, so the electric and magnetic fields are uncoupled (Eq. 2.4). Electric currents and charges are located only on the metallic surfaces, so the fields derive from potentials that are solutions to Laplace's equations. Different techniques are available for their resolution, yielding the electric potential as a function of surface charge density and the magnetic potential as a function of surface current density, from which respectively the capacitances and inductances can be determined.

Static developments cannot take into account the presence of radiation and surface waves. When these effects become significant, a *full-wave* analysis is required, taking into account the six electromagnetic field components.

10.1.3 Open or Closed?

Inherently, circuits are at some point or other enclosed within a metal box and thus become closed structures. On the other hand, antennas are designed to radiate and must remain open to the surrounding space. One would, therefore, expect that techniques designed for boxed-in structures would be used to analyze circuits, keeping those for open structures for antenna analysis.

This is not what happens in practice. The analysis of a structure located within a box requires more parameters than that of an open structure: the specification of the microstrip itself (geometry of the upper conductor, thickness and permittivity of the dielectric) does not suffice, but one must also indicate where and how the structure is located with respect to the enclosure walls, while the box dimensions must also be specified.

As a result, analyses and data are most often reported for open microstrip, with the usual assumption that both the substrate and the ground plane extend to infinity in the transverse direction. Some rules of thumb indicate how far the sides and the cover should be located for the data to be applicable (Section 3.4.10).

The mathematical methods that require a finite space (box) generally place the microstrip structure in the center of the box and later define absorbing boundaries to simulate open space.

10.1.4   Two-Dimensional or Three-Dimensional?

The infinite uniform transmission line is a two-dimensional (2D) structure, since the longitudinal dependence of the fields can be separated from the transverse one. Its analysis only depends on the two transverse coordinates. A finite line, with a beginning and an end, is no longer a 2D structure, but at some distance from the extremities the field structure is not affected by them.

A circular resonator or patch antenna is another 2D structure, in which the axial symmetry allows one to separate the azimuthal dependence from the other two. When the structure is connected to a feed line, it becomes 3D.

All other structures considered are 3D.

10.1.5   Circuit or Antenna?

This distinction is directly related to the planned operation of the structure: a circuit must not radiate, whereas an antenna must radiate as much as possible. Static and quasi-static techniques are applicable to the analysis of circuits, except for spurious effects: radiation and surface waves. Dynamic techniques are, in principle, required for the in-depth analysis of antenna radiation, even though some radiation properties can be derived from simpler models.

10.1.6   Time Domain or Frequency Domain?

Traditionally, dynamic analysis was always conducted in the frequency domain and used the complex notation introduced in Section 2.4.2; the time response, if required, was determined by means of the Fourier transform (Section 2.5).

Nowadays, however, time domain analysis is becoming increasingly popular. It more closely follows the actual physical processes and is thus easier to visualize and understand. In particular, time domain analysis permits the analysis of short-term effects that take place before reflections come back. Its use is most interesting for the idealized lossless models, with constant material properties and perfect conductors.

In the presence of losses, on the other hand, the inherent advantages of time-domain techniques are somewhat reduced. Lossy material properties and boundaries must be taken into account by convolution integrals (Eq. 2.10) [Maloney 1992; Beggs 1992].

10.1.7   A Note on Computing Power

Computer memory and CPU speed experienced dramatic technological advances in the past 20 years. Memory chip density exhibited "a steady annual compound growth rate of more than 70%" since the mid-1970s [Spruth 1989]. It passed the 1 kbit/chip mark in 1974 and is now approaching 1 Gbit/chip. Advances in computer technology opened the new era of *computational electromagnetics*. Previously, electromagnetic problems that did not admit analytical solutions had to be treated approximately by perturbation or variational techniques. Researchers can now access powerful computers and develop sophisticated methods to tackle more complex electromagnetic problems and obtain increasingly accurate solutions.

10.2   CONFORMAL MAPPING

10.2.1   The Electrostatic Problem of Microstrip

In the low-frequency range (Section 10.1.2), or on the cross section of a homogeneous transmission line (Section 3.1.6), the potential is a solution of Laplace's equation. It is constant on the metal boundaries—on the ground plane and on the upper conductor. While the ground plane entirely covers the transverse plane $z = -h$, the upper conductor only partially covers the plane $z = 0$. The boundary conditions are thus transversely inhomogeneous, and Laplace's equation cannot be solved by separation of variables.

10.2.2   Principle of Conformal Mapping

A coordinate system fitting the microstrip boundaries can be defined by conformal mapping, using transformation properties related to complex numbers [Marsden 1973].

A complex number $\underline{z} = x + jy$ is associated with every point in a complex plane, referenced by its rectangular coordinates x and y (the underlined complex number $\underline{z}$ should not be confused with the direction of propagation z). In another complex plane, another point of coordinates u and v is associated with the complex number $\underline{w} = u + jv$. The two points are connected by an analytical complex function (continuous and with a single-valued derivative) that defines the conformal mapping

$$\underline{w} = f(\underline{z}). \tag{10.1}$$

This function maps the transverse plane of the transmission line ($\underline{z}$ plane) onto the complex plane of the transformed function $\underline{w}$. For the transform to be of use, the transformed conductor boundaries must form the four sides of

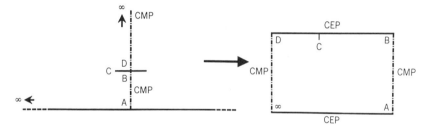

FIGURE 10.1   Conformal mapping for microstrip.

a rectangle (Figure 10.1); hence, the complex geometry of the microstrip is replaced by a finite parallel-plate capacitor.

10.2.3   Limitations

A complex number has real and imaginary parts, which are made to correspond to the two transverse coordinates of a transverse transmission line. Conformal mapping can only be used to analyze 2D structures.

The analytical properties of the transform function are specifically related to Laplace's equations. This means that conformal mapping cannot be extended to consider structures containing a spatial distribution of electrical charges.

Microstrip is inhomogeneous, and conformal mapping replaces the straight air–dielectric boundary by a curve in the transformed structure. Replacing the inhomogeneous dielectric by an equivalent homogeneous one (Section 3.4.1) makes the corresponding capacitor homogeneous and the field within it constant.

10.2.4   Schwartz-Christoffel Transform

The mapping function $f(z)$ providing the desired transformation is determined by a *Schwartz-Christoffel* transform, which "straightens" the angles and provides an integral equation for the function. For a microstrip with an infinitely thin conducting strip, the integration yields a set of simultaneous equations involving theta and elliptic functions [Schneider 1969]. The equations provide the approximate relationships of Section 3.4.2 for the characteristic impedance.

Similar conformal mappings can be established by the Schwartz-Christoffel transform for boxed homogeneous microstrip. While an integral equation can always be written to define the transform, it is not always possible to carry out its integration.

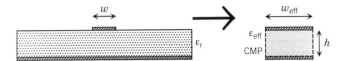

FIGURE 10.2   Equivalent waveguide.

10.3   WAVEGUIDE AND CAVITY METHODS

10.3.1   Equivalent Waveguide Models

The microstrip line is replaced by a rectangular waveguide with two side walls in a perfect magnetic conductor (PMC, Section 2.3.6) homogeneously filled with a dielectric permittivity ε_e (Eqs. 3.17 and 3.42). The width w_{eff} is slightly larger than the width w of the upper conductor, to take into account the fringing fields (Figure 10.2). Its value is obtained from the static analysis of the uniform line (Chapter 3) and can be corrected to account for dispersion [Wolff 1989].

The microstrip problem is replaced in this manner by a rectangular waveguide problem that was solved by Marcuvitz [1951]. This approach was applied to analyze microstrip T-junctions [Leighton and Milnes 1971]. It was then used extensively in Germany to analyze most microstrip discontinuities. The frequency-dependent scattering parameters were determined for the impedance step [Wolff et al. 1972; Kompa 1976a], the bend [Mehran 1975], the mitered bend [Menzel 1976], the T-junction [Wolff et al. 1972; Mehran 1975], the Y-junction [Menzel 1978], and the crossing [Menzel and Wolff 1977]. The LC equivalent circuit of a slot was determined with a simplified version of this model [Hoefer 1977].

10.3.2   Equivalent Cavity Model

In a similar manner, a patch antenna is replaced by an equivalent resonator with a PMC side wall [Lo et al. 1979; Carver and Mink 1981]. The upper patch and the section of the ground plane located below are joined by a magnetic wall under the edge of the patch (Section 11.3.3). The resonator size is slightly enlarged to account for fringing fields. The magnetic currents flowing in the side wall of the cavity radiate when the cavity resonates. It is assumed that the equivalent cavity is surrounded by free space. This model is mostly used for geometries in which the Helmholtz equation possesses an analytical solution, such as disks, rectangles, triangles, or ellipses. It assumes that the substrate thickness is much smaller than a wavelength.

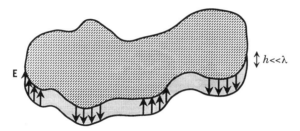

FIGURE 10.3   Planar circuit.

10.3.3 Planar Circuit Analysis

The concept of planar circuit, introduced by Okoshi and Miyoshi [1972], defines a structure in which one dimension is much smaller than the wavelength and where the fields do not vary along it (Figure 10.3). A 2D Helmholtz equation must then be solved, with PMC boundaries except at the locations of the ports where transmission lines are attached. The Green's functions of the problem can be expressed in different manners: integral equation, image theory, contour integrals, eigenfunction expansions, boundary leading to the definition of a generalized impedance matrix. This approach extends the use of the cavity model to the analysis of circuits.

With this approach complex shapes can be analyzed by using segmentation and desegmentation techniques [Sharma and Gupta 1981; Sorrentino 1985]. A complicated shape is divided into simpler elementary shapes, for which the impedance matrices can be calculated, and the parts are connected at equivalent ports to form the original shape (Figure 10.4). This approach was used to analyze many microstrip bends, steps, power dividers, and combiners of various shapes, filters, patch antennas, and arrays. A comprehensive survey was made by Gupta [1989].

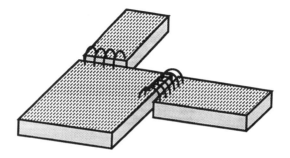

FIGURE 10.4   Segmentation procedure.

10.3.4   Mode-Matching Technique

The mode-matching technique typically considers sharp discontinuities, such as abrupt changes in width. The fields on both sides of the discontinuity are expanded over the normal modes of the corresponding lines, with unknown coefficients. The continuity of the field components across the plane of the discontinuity then provides relationships to determine these coefficients.

Step discontinuities were analyzed in shielded microstrip [Uzunoglu et al. 1988] and in open microstrip lines [Chu et al. 1985]. This approach was also used to determine the effect of conductor thickness [Bögelsack and Wolff 1987].

10.3.5   Transverse Resonance Technique

The transverse resonance technique is similar to the mode-matching technique, but with fields expanded across the guided-wave structure. It was used to characterize printed circuit transmission lines (enclosed in a rectangular cavity) [Yee 1985; Yee and Wu 1986], orthogonally crossed striplines (in a rectangular cavity) [Uwano et al. 1987], shielded microstrip resonators [Jansen 1974], and shielded planar transmission lines [Jansen 1981]. However, when the transverse resonance technique is implemented in the solution of the last two problems, the solution process is basically identical to the integral equation technique (Section 10.6).

10.4   FINITE DIFFERENCES

10.4.1   Basic Principle

The finite-difference technique discretizes derivatives; in other words, it replaces them with finite differences:

$$\frac{\partial F}{\partial u} \to \frac{\Delta F}{\Delta u} \qquad \text{where } \Delta u \text{ remains finite.} \qquad (10.2)$$

The limit $\Delta u \to 0$, which would provide the actual derivative, is not taken, but the size of Δu is kept reasonably small for the approximation to be accurate. A 2D or 3D grid is drawn over the geometry to be analyzed, generally with a square or a cubic mesh with $\Delta u = h$ (Figure 10.5). The field quantities are then considered at all nodes within the mesh and related to other field quantities at the adjacent nodes.

10.4.2   Electrostatic Problem

In the electrostatic problem (Section 10.2.1), Laplace's equation for the potential (Section 3.1.4) is solved in the presence of the boundary conditions

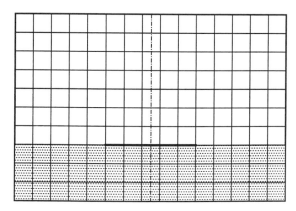

FIGURE 10.5   Square grid for the finite-difference method.

on the conductors. After each derivative is replaced by the corresponding difference, the equation at one node x, y becomes (for a 2D problem)

$$\nabla^2 \Phi \cong \frac{\Phi(x + h, y) + \Phi(x - h, y) + \Phi(x, y + h) + \Phi(x, y - h) - 4\Phi(x, y)}{h^2} = 0.$$

$$(10.3)$$

The potential at point x, y is the average of the potential at the adjacent nodes:

$$\Phi(x, y) = \frac{\Phi(x + h, y) + \Phi(x - h, y) + \Phi(x, y + h) + \Phi(x, y - h)}{4}.$$

$$(10.4)$$

For a 3D problem, one has six adjacent nodes to consider. The application of Eq. 10.4 at every node in the grid that is not on a metal boundary provides a system of N linear equations with N unknowns (N being the number of nodes). This system may be solved by matrix inversion (if N is small enough) or, since the matrix is sparse, by iterative techniques such as relaxation methods [Forsythe and Wasow 1960]. Whenever possible, the boundaries of the conductors should be on grid nodes. Otherwise, one may extrapolate between grid nodes.

More elaborate techniques define a variable grid mesh, providing a finer resolution close to edges and corners, where the field amplitudes are largest. Then the mathematics become more complex. In theory, one might expect to improve the accuracy by decreasing the grid spacing h. This works up to certain point, but rounding errors become significant in large systems of equations, and precautions must be taken in the calculations.

10.4.3 Finite-Difference Time Domain (FDTD)

Finite differences in time domain (FDTD) provides a "full-wave" solution to Maxwell's equations. The method was used in a myriad of electromagnetic problems, including scattering and radar cross section (RCS) calculations, electromagnetic pulse (EMP) penetration problems, transients, inverse problems, and, recently, planar microstrip circuit and antenna problems. With current advances in computer technology (i.e., memory, speed, graphical interfaces), FDTD has become a valuable approach to solve complex electromagnetic problems. Some of the more salient features include the ability to handle various types (even time dependent) of discontinuities within the resolution of the FDTD mesh and the ability to solve transient problems.

Maxwell's curl equations (2.4) are discretized in both the space and time domains [Reineix and Jecko 1989] and thus are replaced by linear equations. The structure and the surrounding space are decomposed into elementary parallelipipedic (or cubic) cells. The electric field components tangential to the cell edges are defined, along with the magnetic field components normal to the cell faces (Figure 10.6). Additional conditions are introduced to take into account dielectric interfaces, metallic boundaries, and wire structures (excitation).

The FDTD method was first introduced to the electromagnetic community by Yee [1966], and detailed descriptions can be found in Taflove and Umashankar [1989]. The time-dependent, source-free, Maxwell curl equa-

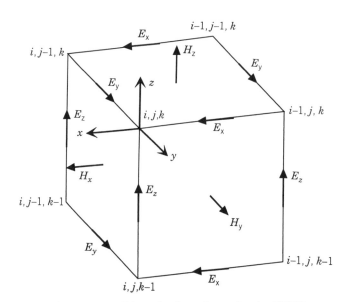

FIGURE 10.6 Discretization scheme for the FDTD.

tions in free space are expressed in terms of the six field components in rectangular coordinates:

$$\mu \frac{\partial H_x}{\partial t} = \frac{\partial E_y}{\partial z} - \frac{\partial E_z}{\partial y}, \qquad \mu \frac{\partial H_y}{\partial t} = \frac{\partial E_z}{\partial x} - \frac{\partial E_x}{\partial z}, \qquad \mu \frac{\partial H_z}{\partial t} = \frac{\partial E_x}{\partial y} - \frac{\partial E_y}{\partial x},$$

$$\varepsilon \frac{\partial E_x}{\partial t} = \frac{\partial H_z}{\partial y} - \frac{\partial H_y}{\partial z}, \qquad \varepsilon \frac{\partial E_y}{\partial t} = \frac{\partial H_x}{\partial z} - \frac{\partial H_z}{\partial x}, \qquad \varepsilon \frac{\partial E_z}{\partial t} = \frac{\partial H_y}{\partial x} - \frac{\partial H_x}{\partial y}.$$

$$(10.5)$$

After discretization in space and time, the time and spatial derivatives are approximated by central differences. The two expressions on the left-hand side in Eq. 10.5 take the form

$$\frac{H_x^{n+1/2}\left(i, j + \frac{1}{2}, k + \frac{1}{2}\right) - H_x^{n-1/2}\left(i, j + \frac{1}{2}, k + \frac{1}{2}\right)}{\Delta t}$$

$$= \frac{E_y^n\left(i, j + \frac{1}{2}, k + 1\right) - E_y^n\left(i, j + \frac{1}{2}, k\right)}{\mu \, \Delta z}$$

$$- \frac{E_z^n\left(i, j + 1, k + \frac{1}{2}\right) - E_z^n\left(i, j, k + \frac{1}{2}\right)}{\mu \, \Delta y},$$

$$\frac{E_x^n\left(i + \frac{1}{2}, j, k\right) - E_x^{n-1}\left(i + \frac{1}{2}, j, k\right)}{\Delta t}$$

$$= \frac{H_z^{n-1/2}\left(i + \frac{1}{2}, j + \frac{1}{2}, k\right) - H_z^{n-1/2}\left(i + \frac{1}{2}, j - \frac{1}{2}, k\right)}{\varepsilon \, \Delta y}$$

$$- \frac{H_y^{n-1/2}\left(i + \frac{1}{2}, j, k + \frac{1}{2}\right) - H_x^{n-1/2}\left(i + \frac{1}{2}, j, k - \frac{1}{2}\right)}{\varepsilon \, \Delta z}. \quad (10.6)$$

Similar expressions are obtained for the four other equations. The electric and magnetic fields are calculated on alternate time steps and depend only on information from the previous time step. The indices (i, j, k) denote position, n denotes the time step, and Δx, Δy, and Δz are the physical distances between nodes in the i, j, and k directions, respectively. Usually $\Delta x = \Delta y = \Delta z$ for all nodes, making the FDTD mesh a cube. Noncubic mesh configurations were also investigated. Time is stepped (or marched) in Δt intervals. To ensure the stability of the method, Taflove and Brodwin [1975a] established a relationship between distances between nodes and the time-step size:

$$v_{max} \, \Delta t \leq \frac{1}{\sqrt{(\Delta x)^{-2} + (\Delta y)^{-2} + (\Delta z)^{-2}}}, \qquad (10.7)$$

where v_{max} is the maximum phase velocity.

The problem to be studied is first discretized by an FDTD mesh. When the structure is finite, as with a closed cavity, the mesh is terminated on the surrounding boundaries. When considering open structures, such as microstrip antennas, one should model infinite space, which is clearly impossible. The structure is then limited by "absorbing sheet" boundaries, made of a ficticious medium with both electric and magnetic losses, defined in such a way that waves reaching the boundary are not reflected [Mur 1981].

Initially, the field quantities are set equal to zero at all nodes, and a time-dependent excitation is applied at an input point into the mesh, for instance, a prescribed field distribution under a microstrip line. The excitation often takes the form of a Gaussian pulse containing a desired frequency spectrum. After its injection into the mesh, time is stepped until all field quantities have decayed to zero. A time-discretized harmonic function can also be injected, in which case time is stepped until all field quantities vary in a time-harmonic manner. This may take several periods if the structure has a large quality factor (Q) [Taflove and Brodwin 1975b].

Output data is extracted at some observation point, where the desired field component is recorded as a function of time. When frequency-domain information is required, a Fourier transform is carried out.

FDTD requires vast amounts of memory and computational time; it only became efficient when large computer memories became available. It was used to determine the scattering of electrically large structures such as aircraft, missiles, and so on. FDTD was also used to analyze microstrip structures: arbitrarily shaped 2D microwave circuits [Gwarek 1988], microstrip in a cavity [Choi and Hoefer 1986], lines, discontinuities, hybrids, and bends [Zhang et al. 1988; Zhang and Mei 1988; Railton and McGeehan 1989; Moore and Ling 1990]. Even microstrip patch antennas have been successfully analyzed [Reineix and Jecko 1989; Sheen et al. 1990].

10.4.4 Transmission Line Matrix

The transmission line matrix (TLM) method similarly replaces the structure by a mesh, either 2D (squares) or 3D (cubes). It was originally developed by Johns and Beurle [1971], who used Huygens' model for plane-wave propagation and defined a set of transmission lines joining adjacent nodes. It presents similarities to the FDTD, since both methods discretize space and time [Hoefer 1989].

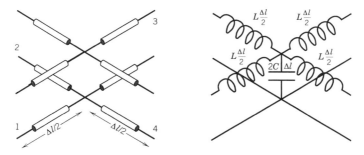

FIGURE 10.7   Shunt-connected transmission lines and equivalent circuit.

The TLM method was first used to solve scattering problems in 2D space, discretized by a mesh of nodes at the junction of two orthogonal TEM transmission lines (Figure 10.7). The voltages and currents on this four-port correspond to the electric and magnetic fields obtained from Maxwell's equations in two dimensions (TE modes).

An incident signal on a port experiences a partial reflection and a partial transmission to the other three ports (Section 6.5.4). A node is represented by a matrix that relates the reflected and incident voltages. Time is stepped to update the voltage at each time interval. The complete structure is modeled by a global matrix, which contains the individual nodal matrices. Boundary conditions define the reflection coefficients at the nodes on the edges. The mesh is excited by voltage impulses introduced at an input node, whereas the output voltage at any node can be recorded as a function of time. Frequency-domain information is then obtained by taking the Fourier transform. The presence of dielectric materials and of losses is simulated by the adjunction of transmission line stubs [Akhtarzad and Johns 1974], while open structures are modeled with equivalent absorbing boundaries. The treatment is carried out in the time domain, and frequency-domain parameters are determined by the Fourier transform.

By combining shunt and series nodes, one can form 3D TLM nodes to model Maxwell's equations in 3D space and time (Figure 10.8).

Three sources of inaccuracy are identified within the TLM method:

1. The propagation velocity depends on the direction of propagation.
2. The finite duration of the input signal (time domain) produces a truncation error.
3. In the vicinity of edges, the field solution is highly irregular: in order to model it accurately, many nodes are required, thereby increasing memory and CPU time.

Techniques for the reduction of errors were introduced [Hoefer and Shih 1980].

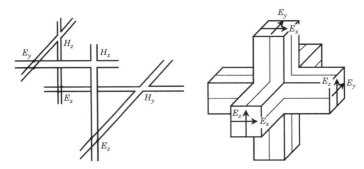

FIGURE 10.8   Three-dimensional TLM nodes: distributed and condensed.

The TLM method was applied to analyze shielded microstrip discontinuities [Johns and Akhtarzad 1975; Akhtarzad and Johns 1975b]. Yoshida and Fukai introduced an alternative way that used gyrators. This approach was applied to analyze a thick microstrip gap [Koike et al. 1985] and a side-coupled microstrip filter [Shibata el al. 1989]. A software package based on the 2D TLM method is commercially available: authored by Hoefer and So (University of Ottawa) it is called *The Electromagnetic Wave Simulator*.

10.4.5   Method of Lines

In the method of lines, the fields are discretized in the transverse directions (tangential to the dielectric interface) but solved exactly for the normal direction. It was used to analyze discontinuities, resonators, and filters [Worm and Pregla 1984].

10.5   VARIATIONAL PRINCIPLES

10.5.1   Basic Description

A variational principle is a mathematical expression involving fields or potentials whose value passes through an extremum (minimum or maximum) when the true value of the fields is introduced. Variational expressions for the capacitance and the inductance are derived from Laplace's equation in the static approximation and for homogeneous transmission lines. Similar expressions are obtained for dynamical cases from Maxwell's equations for propagation factors, resonant frequencies, or terms of the scattering matrix.

Approximations for the fields or potentials are then introduced within the variational principle and adjusted until an extremum is obtained. Several schemes can be used; for instance, an expression for the trial field may be defined with a number of unknown coefficients, and the derivatives with respect to each coefficient is then set equal to zero.

10.5.2   Static Approximation

Solutions of Laplace's equation minimize the electrical energy for specified boundary conditions. When the potential Φ is a solution of Laplace's equations, the quantity

$$F(\Phi) = \int_V |\nabla\Phi|^2 \, dV \qquad (10.8)$$

takes its lowest possible value: any changes in potential Φ within the volume V can only increase the value of F (the potential must always satisfy the boundary conditions). The following strategy can then be used to solve Laplace's equation: consider all possible distributions of the potential Φ satisfying the boundary conditions, and select the one that provides the smallest possible value for F [Harrington 1968].

10.5.3   Galerkin or Rayleigh-Ritz Expansion

The unknown expression for the potential can be expanded over a set of known functions f_i [Morse and Feshbach 1953]:

$$\Phi_a = \sum_{i=1}^{M} c_i f_i. \qquad (10.9)$$

The functions $f_1, f_2, \ldots, f_M$, with M tending to infinity, form a complete set of independent functions, and the coefficients c_i become the unknowns that must be determined. In principle, any field distribution can be expressed exactly as a sum of functions of the set, and the development is unique. For practical reasons, however, the set must be truncated; that is, M must remain a finite number so that computations can be carried out: this introduces an approximation. The functions can be defined over the complete structure under study, for instance in the form of Fourier series, or only over limited subsections of the structure.

The expansion (Eq. 10.9) is introduced within Eq. 10.8, and the extrema of $F(\Phi_a)$ are determined by setting the derivatives equal to zero:

$$\partial F(\Phi_a)/\partial c_i = 0. \qquad (10.10)$$

In this fashion, one obtains a set of M linear equations with M unknowns that is solved for the coefficients c_i. A good estimate of the true value of F may be obtained with a rough estimate for the potential distribution Φ_a. The convergence is determined by increasing the number M until the value calculated for F tends to a limit (Figure 10.9).

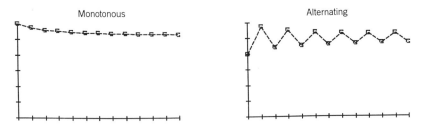

FIGURE 10.9 Determination of the convergence when using a truncated sequence of functions.

10.5.4 Finite Elements

The finite-element method (FEM) was used by civil and mechanical engineers to solve problems in structural, heat, mass-transfer, elasticity, and fluid mechanics problems. It was introduced to electrical engineering around the late 1960s to analyze propagation in dielectrically loaded metallic waveguides with arbitrary cross sections. [Strang and Fix 1973]. The structure to be analyzed is subdivided into small cells—triangles for the 2D case (Figure 10.10) and tetrahedra in the 3D case.

The unknown potential Φ within each region is approximated by a polynomial expansion in terms of the coordinates x, y, and z. A first-order polynomial provides the simplest possible choice, providing in the 2D case,

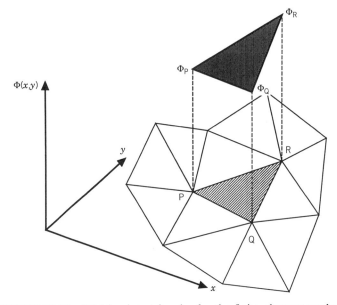

FIGURE 10.10 Division into triangles for the finite-element method.

within a triangle having corners P, Q, and R (Figure 10.10),

$$\Phi(x, y) = Lx + My + N. \tag{10.11}$$

The coefficients L, M, and N are directly related to the potentials on the three corners of the triangle. The first two terms are

$$L = \frac{\Phi_P(y_Q - y_R) + \Phi_Q(y_R - y_P) + \Phi_R(y_P - y_Q)}{x_P(y_Q - y_R) + x_Q(y_R - y_P) + x_R(y_P - y_Q)},$$

$$M = \frac{\Phi_P(x_Q - x_R) + \Phi_Q(x_R - x_P) + \Phi_R(x_P - x_Q)}{y_P(x_Q - x_R) + y_Q(x_R - x_P) + y_R(x_P - x_Q)}. \tag{10.12}$$

The value of N is not determined because it does not appear in subsequent developments. The contribution of this particular triangle to the function F(Φ) in Eq. 10.5 is then

$$(L^2 + M^2)S_{PQR}, \tag{10.13}$$

where S_{PQR} is the area of the triangle. Summing the contributions of all the triangles within the cross section yields an approximation for F(Φ) that is a function of the potentials Φ_I at the corners of all the triangles. Since the smallest possible value for this function must be determined, its derivative with respect to every one of the Φ_I's is set equal to zero:

$$\frac{\partial F(\Phi_A, \Phi_B, \dots, \Phi_Z)}{\partial \Phi_I} = 0. \tag{10.14}$$

This provides a system of N linear equations with N unknowns, N being the number of corners of triangles not located on a metal boundary. Since the potential at any point appears only within the expressions from the triangles adjacent to this point, the corresponding matrix is sparse. The system may be solved by matrix inversion or other suitable algorithms.

10.5.5 Extension to the Dynamical Case

A vector variational formulation can be expressed in terms of the magnetic field and is used in 3D problems:

$$F(\underline{H}) = \int_V \left(\frac{1}{\varepsilon}(\nabla \times \underline{H}) \cdot (\nabla \times \underline{H})^* - \omega^2 \mu \underline{H} \cdot \underline{H}^* \right) dV. \tag{10.15}$$

Equation 1.15 is stationary with respect to perturbations about the true magnetic field solution and can be related to the stored energy. The entire surface or volume is then divided into finite elements, generally triangles for a surface or tetrahedrons for a volume (Figure 10.10). The field is expressed by polynomial expansions within each element. The functional is then minimized with respect to the polynomials. This produces a sparsely populated matrix. When dealing with a large number of unknowns, one can employ special sparse matrix solvers, thereby saving considerable amounts of memory and CPU time [Davies 1989].

Modal behavior in shielded microstrip lines was investigated [Daly 1971]. Various algebraic functions were considered to approximate the vector field over a 2D finite element [Konrad 1977]. Attention began to shift toward 3D problems. It soon became apparent that the FEM formulation was plagued by serious problems: singularities for longitudinal field formulation, "spurious solutions" when using vector formulations. Many FEM papers were devoted to these problems, and many different methods were suggested to eliminate spurious solutions: penalty function to enforce a divergenceless field, expansion over divergence-free fields, special finite elements to confine spurious solutions to a particular frequency range, discrimination techniques to separate "real" from "spurious" solutions, six-field-component formulation, modifications to the governing vector equations, tangential finite elements.

FEM has been used in the quasi-TEM analysis of transmission lines [Pantic and Mittra 1986]. Skin effect in quasi-TEM analysis has also been accounted for [Costache 1987]. In an interesting approach, Csendes and Lee combined the planar waveguide model of microstrip circuits and the transfinite element method to predict the performance of practical MMIC structures [Csendes and Lee 1988]. Hewlett–Packard released a software package named the High-Frequency Structure Simulator (HFSS), developed at Anasoft Corporation [Csendes 1991]. HFSS models complicated 3D passive microwave circuits with FEM to determine their scattering parameters and field distributions.

10.6 INTEGRAL EQUATION TECHNIQUES

Integral equation formulations for the analysis of microstrip structures are the most elaborate techniques available and provide "almost rigorous" solutions. They are also the most computer intensive, requiring considerable computation time and computer memory. They can be used as "standard references," providing data against which results provided by approximations may be validated. Nevertheless, with the increasing availability of computing power, they have become quite popular for analyzing microstrip antennas and simple circuits. Described briefly in this chapter are the general principles underlying the theory and a succinct account of some commonly found numerical implementations.

10.6.1   Sources, Boundaries, and Green's Functions

An integral equation description for a microstrip circuit or an antenna is obtained by specifying the boundary conditions that must be satisfied by the electromagnetic field. The conditions apply to the total fields, formed by an excitation (impressed field) and a scattered part (induced field). The scattered field is produced by a set of unknown induced sources (currents and charges on the conductors). Within any linear system, the scattered field (or the potentials) can be expressed as a superposition integral of the unknown sources, weighted by Green's functions, that are the fields (or the potentials) produced by elementary point sources:

$$\underline{\mathbf{A}}(\mathbf{r}) = \int_{V'} \overline{\overline{\mathbf{G}}}_A(\mathbf{r}|\mathbf{r}') \cdot \underline{\mathbf{J}}(\mathbf{r}') \, \mathrm{d}v', \qquad (10.16)$$

$$\underline{V}(\mathbf{r}) = \int_{V'} \underline{G}_V(\mathbf{r}|\mathbf{r}') \underline{q}(\mathbf{r}') \, \mathrm{d}v'. \qquad (10.17)$$

These expressions represent the potentials at the point $\mathbf{r}$, the integration being carried out over the variable $\mathbf{r}'$. The integral is evaluated over the volume containing the sources. Introducing the integral formulation for the scattered field into the boundary conditions yields a set of integral equations for the unknown induced sources.

When formulating an integral expression, the essential step is the determination of the scattered fields produced by the induced currents. With the Green's function concept, this is given by Eqs. 10.16 and 10.17. The same goal can also be achieved without Green's functions, replacing them by modal waveguide expansions (Section 10.6.6), spheroidal harmonics [Hechtman et al. 1991], complex images [Alanen et al. 1986], Floquet modes, Green's theorems, or other mathematically equivalent devices.

By applying boundary conditions on the dielectric interface and on all metallic surfaces (ground plane and upper conductors), a set of coupled integral equations is obtained. One or several boundary conditions can be included in the Green's functions, thus reducing the number of integral equations at the cost of greater complexity of the Green's functions. When a modified Green's function satisfies some boundary conditions, any field obtained as a superposition integral with this Green's function satisfies the same boundary conditions.

The formulation of the problem can also be modified by defining "induced sources." The surface equivalence theorem specifies that, in electromagnetics, a given set of sources can be replaced by another set of sources defined on a surface enclosing the original set. A dielectric layer can also be replaced by a distribution of polarization currents in free space (volume equivalence theorem). Many different integral equation models were defined, depending on how boundary conditions are considered: they can be

Used to generate an integral equation
Included in the Green's function definition
Replaced by unknown equivalent sources

10.6.2 Two-Step Operation

The resolution of the problem must be carried out in two steps:

1. Determination of the Green's function or alternative formulation for the fields produced by a point source (Sections 10.6.3–10.6.6)
2. Determination of the surface current and charge distributions that meet the boundary conditions (Eq. 10.21) and also correspond to the specified excitation (Sections 10.6.7 and 10.6.8)

10.6.3 Classification of Integral Equation Models

Since many criteria can be used to classify integral equation models, it is practically impossible to produce a rigorous and coherent classification scheme encompassing all existing versions. Three main groups can be defined:

1. Homogeneous Green's functions (Section 10.6.4)
2. Stratified media Green's functions (Section 10.6.5)
3. Waveguide modal expansions (Section 10.6.6)

10.6.4 Homogeneous Green's Functions

Integral equations can be defined with the simplest possible Green's functions, defined for free space (Section 2.6.8):

$$\underline{\mathbf{A}}(\mathbf{r}) = \frac{\mu_0}{4\pi} \int_{V'} dV \frac{\underline{\mathbf{J}}(\mathbf{r}') \exp\left(-j\omega\sqrt{\underline{\varepsilon}\underline{\mu}}\,|\mathbf{r} - \mathbf{r}'|\right)}{|\mathbf{r} - \mathbf{r}'|}$$

$$= \int_{V'} dV\, \underline{\mathbf{J}}(\mathbf{r}') G_A(\mathbf{r} - \mathbf{r}'). \tag{2.45}$$

The dielectric substrate is replaced by unknown polarization currents, which must be numerically determined, together with the unknown conduction surface currents on the metallic surfaces [Kishk and Shafai 1986a]. Another approach replaces the dielectric layers by magnetic and electric equivalent surface currents [Kishk and Shafai 1986b; Sarkar and Arvas 1990]. In general, the use of free-space Green's functions leads to problems of great numerical complexity, with very large linear systems of equations. For boxed configurations the lateral conducting walls must also be replaced by equivalent surface currents, increasing even further the final number of unknowns.

10.6.5 Stratified Media Green's Function

At the beginning of the twentieth century, the German physicist Arnold Sommerfeld [1909] formulated rigorous expressions for the fields produced by a wire above a lossy ground. By an extension of his approach, the boundary conditions associated with the dielectric substrate and the ground plane are included in the Green's functions, which become Sommerfeld integrals. The sources are the electric surface currents and charges on the upper conductor, and it is generally assumed that the dielectric slab and the ground plane extend to infinity in the transverse directions.

The functions that must be integrated possess singularities and sometimes exhibit a diverging behavior for increasing values of the argument (Figure 10.11). The evaluation of Sommerfeld integrals thus presented severe numerical problems, until Mosig [1988] developed efficient computer algorithms for their accurate evaluation.

This approach includes many different versions, among them the classical electric field integral equation [Das and Pozar 1989], the spectral-domain approach [Itoh 1980], the reaction theorem technique [Newman and Tulyathan 1981], the Pocklington equation [Katehi and Alexopoulos 1984], static approximations [Farrar and Adams 1972; Silvester and Benedek 1973], and the mixed potential integral equation [Mosig and Sarkar 1986]. Microstrip circuits and discontinuities were analyzed [Katehi and Alexopoulos 1985; Koster and Jansen 1986; Skrivervik and Mosig 1990; Hill 1991; Wu

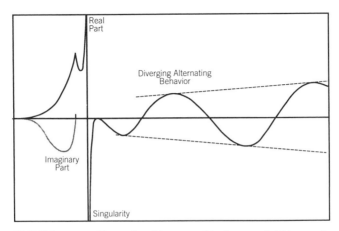

FIGURE 10.11 Example of integrand in Sommerfeld integral.

et al. 1991]. Practical implementations are commercially available as software packages such as LINMIC [Jansen 1988].

Extensions to covered problems are possible with a slight modification in the Green's functions [Yang et al. 1992]. For boxed problems, however, the Green's functions must be fully redefined in order to include the boundary conditions associated with the box.

10.6.6 Waveguide Modal Expansions

When one specifically considers boxed-in microstrips, it is interesting to expand the electromagnetic field over waveguide modes. The box is considered as two cascaded waveguide sections, both filled with a homogeneous dielectric. The upper conductor is located at the interface of the waveguide sections. Boundary conditions at the interface are applied on the waveguide field expansions, thus avoiding the explicit use of a Green's function. This approach was used to analyze open-ended microstrip lines [Rautio and Harrington 1987], a four-resonator coupled line filter [Dunleavy and Katehi 1988], and microstrip line discontinuities [Hill and Tripathi 1991]. Two software packages use this approach to analyze boxed-in microstrip discontinuities: Microwave Explorer by Compact Software Inc. [Hill et al. 1992, Section 12.3.1] and EM by Sonnet Software Inc. (Section 12.3.5).

10.6.7 Boundary Conditions

The surface currents flowing on the conductors and the corresponding surface charge distribution are not yet known and must be determined by imposing the resistive boundary conditions for the tangential electrical field on the surface S' of the upper conductor [Mosig et al. 1989]:

$$\mathbf{e}_z \times \{\underline{\mathbf{E}}^e(\rho) + \underline{\mathbf{E}}^s(\rho)\} = \underline{Z}_m \mathbf{e}_z \times \underline{\mathbf{J}}_s(\rho), \qquad (10.18)$$

where $\mathbf{E}^e$ is the excitation field, $\mathbf{J}_s$ is the surface current density on S' induced by $\mathbf{E}^e$, $\mathbf{E}^s$ is the scattered field radiated by $\mathbf{J}_s$, and Z_m is the surface impedance on S', taking the finite conductivity of metallic sheets into account.

$$\underline{Z}_m = \sqrt{\frac{j\omega\mu}{\sigma}} = (1 + j)\sqrt{\frac{\omega\mu}{2\sigma}} = (1 + j)R_m, \qquad (2.43)$$

where ω is the operating angular frequency, μ is the metal permeability, and σ is the effective conductivity of the metal, including the possible effect of roughness. The surface impedance $\underline{Z}_m$ vanishes on a perfect electric conductor.

When potentials are considered, Eq. 10.18 becomes

$$\mathbf{e}_z \times \underline{\mathbf{E}}^e(\rho) = \mathbf{e}_z \times \{j\omega\underline{\mathbf{A}}(\rho) + \nabla\underline{V}(\rho) + \underline{Z}_m\underline{\mathbf{J}}_s(\rho)\}, \qquad (10.19)$$

where $\underline{\mathbf{A}}$ and $\underline{V}$ are related through the Lorentz gauge. The potentials are then replaced by their expressions from Eqs. 10.16 and 10.17, yielding the mixed potential integral equation (MPIE) formulation

$$\mathbf{e}_z \times \underline{\mathbf{E}}^e(\rho) = \mathbf{e}_z \times \left\{ j\omega \int_{S'} \overline{\overline{\mathbf{G}}}_A(\rho|\rho') \cdot \underline{\mathbf{J}}_s(\rho') \, ds' \right.$$
$$\left. + \nabla \int_{S'} \mathbf{G}_V(\rho|\rho')\underline{q}_s(\rho') \, ds' + \underline{Z}_m\underline{\mathbf{J}}_s(\rho') \right\}. \quad (10.20)$$

In this expression, the fields and the potentials are determined on the upper conductor at the point $\mathbf{r} = \rho$, and are produced by surface charges q_s and currents $\underline{\mathbf{J}}_s$ at the point $\mathbf{r} = \rho'$ on the surface of the upper conductor (related through the continuity equation).

10.6.8 Method of Moments

A technique of great flexibility and numerical stability is required to solve Eq. 10.20 on the irregular geometries of microstrip circuits. A method of moments (MoM) with subsectional basis and testing functions is generally the most suitable [Harrington 1968]. Overlapping rooftops can be used to develop the currents in the x and y directions, yielding pulse doublets for the surface charge density. The elementary rectangular cells of the division can take various sizes, so any kind of microstrip structure can be modeled [Skrivervik and Mosig 1990] (Figure 10.12).

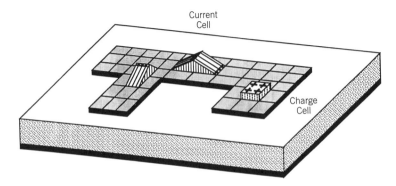

FIGURE 10.12 Definition of subsectional basis functions.

The current and charge density are given by

$$\mathbf{J}_s = \sum_{i=1}^{n} \underline{\alpha}_i T_i^x \mathbf{e}_x + \sum_{j=1}^{m} \underline{\alpha}_j T_j^y \mathbf{e}_y,$$

$$j\omega \underline{q}_s = \sum_{i=1}^{n} \underline{\alpha}_i \Pi_i^x + \sum_{j=1}^{m} \underline{\alpha}_j \Pi_j^y, \tag{10.21}$$

where n and m are, respectively, the number of x- and y-directed basis functions, T_i^x and T_i^y and x- and y-directed rooftop functions, and the corresponding pulse doublets are

$$T_i^s = \begin{cases} \dfrac{1}{w_i}\left(1 - \dfrac{s}{L_i^+}\right), \\[2mm] \dfrac{1}{w_i}\left(\dfrac{s}{L_i^-} - 1\right), \\[2mm] 0, \end{cases} \qquad \Pi_i^s = \begin{cases} \dfrac{1}{w_i L_i^+} & \text{for } s_i < s < s_i + L_i^+, \\[2mm] \dfrac{-1}{w_i L_i^-} & \text{for } s_i - L_i^- < s < s_i, \quad s = x, y, \\[2mm] 0 & \text{elsewhere.} \end{cases}$$

$$\tag{10.22}$$

A Galerkin testing procedure is then applied to satisfy the boundary conditions (Eq. 10.20). The final MoM matrix equation to be solved is then written symbolically as

$$\left[\underline{Z}^{\text{MoM}}\right][\underline{\alpha}] = [\underline{V}], \tag{10.23}$$

where the moment matrix $\underline{Z}^{\text{MoM}}$ is obtained from Eq. 10.20 and the vector $[\underline{V}]$ is directly related to the excitation model. This matrix equation is then solved, yielding the current and charge distribution on the upper conductor of the microstrip, from which the quantities of interest in circuits (impedance or scattering matrix) and in antennas (gain, radiation pattern) can be determined.

10.6.9   Modeling of the Feed

In antenna analysis, a coaxial feed model is most often used (Figure 10.13), because it allows direct comparison between measurement and simulation. The simplest coaxial feed model is a vertical filament of unit current terminated by a point charge acting on some point of the upper conductor.

This simple model, however, does not permit the computation of the self term of the input impedance, because the charge distribution is singular. To do this, one introduces an attachment mode between the coaxial pin and the

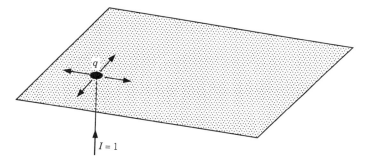

FIGURE 10.13   Coaxial feed model.

upper conductor. The pin is located at the center of a basis cell and the current density is spread over the cell. Since the attachment mode becomes an additional unknown in the MoM procedure, its numerical complexity is increased [Hall and Mosig 1989]. It is possible to avoid undue complexity by modeling the coaxial probe by a vertical current filament of unit density acting on the center of an elementary cell and spreading the associated point charge over the entire cell with a constant density. The continuity equation is not satisfied locally but only globally over the complete cell.

The input impedance is then obtained by integrating the electric field tangent to the vertical filament between the ground plane and the upper conductor. The application of the reciprocity principle then yields the impedance matrix

$$\underline{Z}_{ij} = [\underline{\alpha}_j]^{\mathrm{T}}[\underline{V}_j] = [\underline{V}_i]^{\mathrm{T}}[\underline{Z}^{\mathrm{MoM}}]^{-1}[\underline{V}_j]. \tag{10.24}$$

This formulation for the excitation and impedance leads to a quadratic form for the computation of the impedance matrix. This provides good stability and fast convergence of the numerical process.

10.6.10   Spectral-Domain Approach

The currents on the upper conductor can be expressed by a Fourier transform, and subsequent developments can be carried out in the transformed domain. In this manner, one avoid the (previously) difficult integration of badly behaved functions (Figure 10.11) required to obtain the Green's function in the spatial domain. This method was used to analyze dispersion on microstrip lines [Itoh and Mittra 1973b], impedance steps [Koster and Jansen 1986], and antennas [Araki and Itoh 1981]. It is restricted to simple geometries.

10.7   REMARKS AND COMMENTS

10.7.1   Numerical Aspects

Many techniques used for the analysis of microstrip circuits require the resolution of large systems of linear equations. This can be done by matrix inversion when the number of unknowns is not too large. Various algorithms are available for this purpose in standard software packages.

For very large numbers of unknowns, matrix inversion requires a very large memory space and becomes impractical or extremely time consuming. One may then take advantage of the fact that matrices are sparse (i.e., with many zeros) to implement iterative techniques like overrelaxation or conjugate gradient methods. Also, some terms in the matrices become negligibly small and may safely be set equal to zero.

10.7.2   Simple and Complex Methods

For low-frequency applications, approximate methods generally provide sufficient information. Their main advantage lies in their simplicity, resulting in a relatively small computational effort, and they allow easy synthesis of microstrip structures. Hoffmann [1987] summarized all these techniques and applications with extensive references in his comprehensive book.

However, microstrip structures are used in high-frequency bands where their transverse dimensions and substrate thickness become significant fractions of a wavelength. The equivalent circuits of microstrip discontinuities (steps, bends, etc.) based on a quasi-static approach are no longer adequate; hence, full-wave models are required to analyze effects not included in the TEM approximations. These models just take into account the dispersion and coupling produced by radiation and surface waves. It was also pointed out that, in some situations, the extrapolation toward low frequencies of high-frequency results sometimes led to erroneous values [Easter 1975].

CHAPTER ELEVEN

Microstrip Antennas

Microstrip antennas provide interesting features for aerospace applications, in particular their low weight and thin profile. By combining patches into arrays, one can overcome their inherently low directivity. They are easily mounted on flat or gently curved surfaces.

11.1 GENERAL BACKGROUND

11.1.1 Description

Printed antennas (patches, microstrip) use square, rectangular, circular, triangular, elliptical, or even more complex shapes, as radiating elements. Shape selection depends on the parameters to be optimized: bandwidth, side lobes, cross-polarization. Since patches only radiate close to their resonant frequencies (Section 7.1), their main dimension is about a half-wavelength. The directivity of a microstrip patch is therefore comparable to that of a half-wave dipole (i.e., quite low). This drawback may, however, be overcome by grouping a number of patches to form an array. High-directivity microstrip antennas are realized in this fashion [Pozar 1985a].

11.1.2 History

Although Deschamps introduced the concept of printed antennas in 1953, more than 20 years went by before the first actual realization [Munson 1974]. Considerable interest in microstrip antennas then quickly developed, as evidenced by an important specialist's meeting held in Las Cruces, New Mexico [Carver 1979], which was in turn followed by publications [Chang 1981] and books [James et al. 1981; Bahl and Bhartia 1980].

Microstrip antennas were first made with printed circuit substrates, which are thin and have a relatively high permittivity. They exhibited, therefore, a

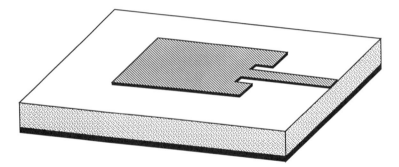

FIGURE 11.1   Patch antenna with microstrip line feed.

narrow bandwidth of around a few percent [Derneryd 1988]. This significantly reduces their interest for many practical applications. On the other hand, simple mathematical models could be used for their analysis (Section 11.3).

These drawbacks are avoided by building antennas on thick low-permittivity substrates, where larger bandwidth and efficiency are achieved. A more rigorous analysis of the field distribution within the structure then becomes necessary, leading to the development of integral equation techniques (Section 10.6).

11.2   ANTENNA FEEDS

11.2.1   Transmission Line Feed

The simplest way to feed a microstrip patch is undoubtedly to combine the microstrip line feed and the microstrip patch on the same substrate (Figure 11.1). However, this structure cannot be optimized either as an antenna or as a transmission line, because the specific requirements for both are contradictory (Section 1.3.6). This means that some compromise between the two must be made, in which the feed line does not radiate too much at the discontinuities [Hall and Hall 1988]. The spurious radiation increases the side-lobe level and the cross-polarization, downgrading the antenna performance. On the other hand, considerable reactive power is accumulated below the patch (cavity effect), reducing its bandwidth.

11.2.2   Coaxial Feed

Theoretical developments nearly always consider coaxial feeds located perpendicularly to the patch antenna (Figure 11.2). Theoretical models were developed to study the current injection into the patch (Section 10.6.9). The

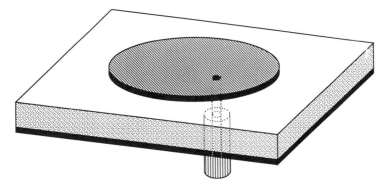

FIGURE 11.2   Patch antenna with coaxial line feed.

intrinsic radiation of the coaxial feed is small and can be neglected when the substrate is thin, but it becomes significant for thick substrates, in which case the feeds of adjacent elements can also couple to one another.

In practice, however, coaxial feeds are difficult to realize, because holes have to be drilled or punched through the substrate, an operation that one generally tries to avoid (Section 13.6.7). Conductors must be introduced through the holes and soldered to the patch, which requires careful handling, and mechanical control of the connection is often difficult. The feed circuit can be placed on the underside of the antenna and shielded by the ground plane.

11.2.3   Buried Feed

Some amount of decoupling between antenna and feed is obtained by placing the patch and the feed at different levels [Lepeltier et al. 1985]. The radiation from the feed can be reduced by using two different substrates with $\varepsilon_{r2} < \varepsilon_{r1}$ (Figure 11.3). Simple models are not suitable for analyzing multilayer structures.

11.2.4   Slot Feed

A more complete separation between radiation and feed is obtained by using the ground plane to shield the radiating patch from the feed system. Coupling between the two is provided by a slot [Pozar 1985b]. The radiation from the line is then physically separated from that of the patch and can be completely avoided by enclosing the feed within a box. To avoid radiation to the back of the antenna the slot must not resonate within the operating frequency band and should be placed far enough from the edge of the patch.

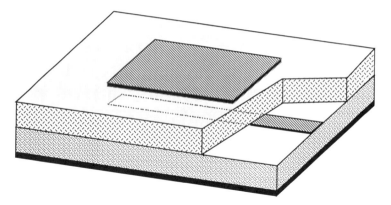

FIGURE 11.3   Patch antenna with buried feed.

11.3   SIMPLE MODELS

11.3.1   Equivalence Principle

The radiation of microstrip antennas is provided by surface electric currents on the upper conductor. However, looking at the structure from above, the radiation seems to come from under the edges of the upper conductor and their close vicinity (Figure 1.5). Assuming that the fields under the upper conductor are the same as in a magnetic wall resonator, the equivalence principle may be used, and magnetic currents introduced as sources along the side walls. Under this assumption, the fields of both models can become equivalent by using Green's functions that are specific to the problem considered [Chuang et al. 1980]. Simple antenna models then replace the patch by radiating apertures, located either on the plane of the interface or between the two conductors and perpendicular to them. Many publications simply consider the electric or magnetic surface currents as sources of radiation in free space, an approximation that may be sufficient for first-order analysis of the far-field pattern of thin substrate antennas.

11.3.2   Transmission Line Model

Since radiation comes out from the patch edges, some authors replace them by radiating apertures fed by a section of transmission line [Munson 1974; Derneryd 1978]. A rectangular patch antenna is then replaced by two vertical radiating apertures under the open-ended edges of the upper conductor. The magnetic current in the apertures is assumed to be constant, and the equivalent conductance of the aperture is determined in terms of the open parallel-plate waveguide [Bahl and Bhartia 1980]. The two apertures are connected by a transmission line section one half-wavelength long (consider-

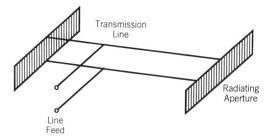

FIGURE 11.4   Transmission line model.

ing the effective wavelength of the open microstrip line, Section 11.4). The input resistance is then a function of the distance from the edge of the patch to the feed point [Derneryd 1978]. The transmission line model (Figure 11.4) was later improved to take into account substrate and conductor losses, coupling between the two apertures, and reactive effects [Dubost 1981; Lier 1982]. This approach was also extended to consider radiation from circular or annular patches.

11.3.3   Cavity Model

The upper patch and the section of the ground plane located below it joined by a magnetic wall under the edge of the patch form a dielectric resonator (Section 7.3 Figure 11.5). The magnetic currents flowing on the cavity side walls radiate at the resonant frequencies of the cavity [Lo et al. 1979; Carver and Mink 1981], which is assumed to be surrounded by free space. The radiation resistance is determined from the integrated power in the far field and the input impedance from cavity theory [Gardiol 1984]. This model is best suited for geometries in which the Helmholtz equation possesses an analytical solution, such as disks, rectangles, triangles, or ellipses. More complex shapes can then be analyzed by segmentation and desegmentation techniques (Section 10.3.3).

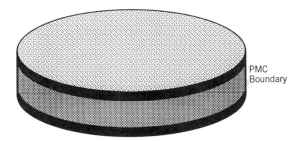

FIGURE 11.5   Cavity model.

11.3.4 Limitations of the Simple Models

Simple models are applicable when substrates are thin, typically less than 0.1λ thick. The ground plane acts as a mirror, and the fields radiated by two parallel plates close to each other tend to cancel out. The radiation efficiency then remains rather low, with a narrow frequency band (close vicinity of resonances).

Simple models provide a first-order approximation of the radiation pattern of the antenna but do not take surface waves into account [James et al. 1981]. A finer analysis requires the use of the integral formulation in terms of Green's functions (Sommerfeld integrals) and a numerical resolution on the computer in terms of a moment method [Mosig and Gardiol 1982; Mosig et al. 1989].

11.4 BROAD-BAND ANTENNAS

11.4.1 The Folded Dipole Antenna

The antenna bandwidth can be widened by combining several resonant modes, with resonance frequencies close to one another, as in the symmetrical folded dipole antenna (Figure 11.6) [Dubost 1981]. This approach tends to produce a frequency-dependent squint angle that may become troublesome when antennas are grouped to form arrays.

11.4.2 Parasitic Elements

The bandwidth can also be increased by adding parasitic elements with slightly different sizes next to the main radiator (Figure 11.7) [Aanandan and Nair 1986]. By adjusting the dimensions, it is then possible to maintain an almost constant radiation pattern (the measured near fields for this structure are shown in Figure 14.8).

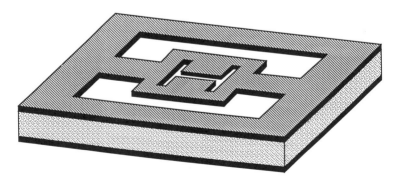

FIGURE 11.6 Folded dipole antenna.

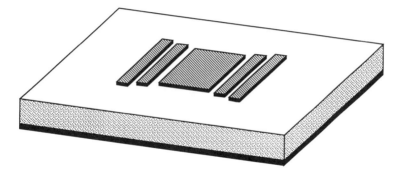

FIGURE 11.7   Patch antenna with parasitic elements.

11.4.3   The SSFIP Principle

To radiate properly, an antenna should be built on a thick substrate with low permittivity. Ideally, the dielectric substrate should be air, with $\varepsilon_r = 1$, but this is not practical for mechanical reasons. Dielectric foams with a relative permittivity as low as $\varepsilon_r = 1.07$, and good mechanical properties are now available. The use of a thick low-permittivity substrate prevents unwanted concentration of the fields within the substrate; hence, the resonance is broader and the operating bandwidth of the antenna is widened. Surface waves are not significantly excited on low-permittivity substrates, so the coupling between elements in an array is small and may be neglected. Metal patches cannot be deposited directly onto the foam and are therefore deposited on the underside of a thin plastic sheet, which also serves as a protective cover against moisture and environmental effects (radome).

Since requirements for lines and antennas are incompatible, the two functions are clearly separated: the feed is located below the patch on the other side of the ground plane. Coupling from the line to the patch is made through a slot which is realized during the photolithographic process. The slot itself must not resonate over the operating frequency band of the antenna because that would produce back radiation. On the underside of the ground plane, the feed line is deposited on a regular microstrip substrate. The resulting sandwich structure is called a strip-slot-foam-inverted patch antenna (SSFIP), indicating the sequence of features seen by the radiated signal (Figure 11.8) [Zürcher 1988].

Many adjustable parameters are available to optimize the design: substrate permittivity, feed line thickness, foam thickness, coupling slot size, cover material [Zürcher et al. 1989]. A thorough analysis was carried out, providing a 13.2% frequency bandwidth at a 2:1 VSWR, and a radiation pattern quite stable with frequency. The cross-polarization is quite low, between -25 and -20 dB. The measured gain ranges over 8.2–8.8 dB, it is slightly larger than the gain for a regular microstrip patch.

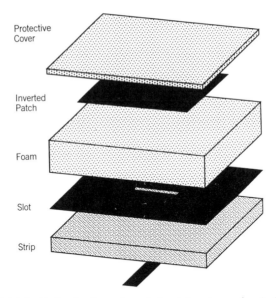

FIGURE 11.8 Strip-slot-foam-inverted patch antenna (exploded view).

Stacking an additional layer of foam and a second patch on top of the first one allowed a frequency bandwidth of 33% to be obtained for a 2 : 1 VSWR. The radiation does not significantly vary over the extended band, and the cross-polarization level is larger but still mostly below the −20-dB level. The gain remains in the same range as that for a single-patch SSFIP.

The latest developments include a "bow tie" slot to improve coupling and parasitic elements to widen the band. Figure 11.9 shows the reflection measured on an antenna for the 1.7–1.9 GHz band.

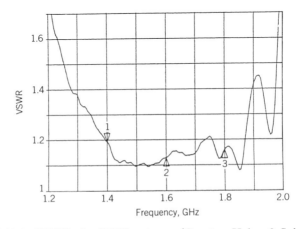

FIGURE 11.9 VSWR of a SSFIP antenna. (Courtesy Huber & Suhner Co.)

The SSFIP concept optimizes the electrical and mechanical characteristics of planar antennas and incorporates a protective radome. SSFIP antennas exhibit high gain, large bandwidth, good mechanical characteristics, low weight, and cost-effectiveness. The components are designed by the photolithographic process. There is no need to drill holes through substrates and ground planes to connect pins or coaxial lines. Because there are no connections between different metal layers, the intermodulation level is reduced. SSFIP antennas are flat and inconspicuous—they can be mounted on building walls and in places where other antenna types would be objectionable.

11.5 PERIODIC ARRAYS

11.5.1 Description

Applications of microwave antennas in communications and radar nearly always require highly directive antennas. Microstrip patches have sizes around $\lambda/2$ and are thus low-directivity radiators. For this reason, they are seldom used alone but are grouped in arrays, with the radiating fields of the elements adding in phase in the main beam and canceling in other directions. Depending on the geometric disposition of the elements, one obtains rectangular, circular, or triangular arrays, or arrays formed by clusters or subarrays of elements. A rectangular array is shown in Figure 11.10.

The fields radiated by the array are obtained by adding the fields radiated by all the patches while taking the interactions between the patches into account [Bhattacharyya and Shafai 1986]. The radiation pattern and the input impedance of a patch are affected by the surrounding elements and, therefore, depend on the position of the patch within the array. A first approximation can be obtained simply by summing the contributions of single patches while neglecting interactions, although this may not be accurate enough in some situations, particularly for phased arrays (Section 11.5.4).

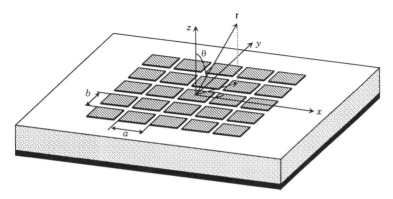

FIGURE 11.10   Rectangular array of microstrip patch radiators.

Adjacent patches must be located close enough to each other for radiation to concentrate in only one direction. When elements are fed in phase (broadside array), the spatial period of the array should not exceed one wavelength. It must be further reduced as the phase difference between elements increases (in phased arrays). Otherwise, spurious "grating lobes" appear in addition to the main beam.

If the array were infinite in size, the contributions from all patches would add in a single direction and completely cancel everywhere else. But in practice an array always has a finite size, so the canceling effect is not complete. The main beam has a finite width and is surrounded by side lobes. The beamwidth and the side-lobe power ratio determine the size of the array.

11.5.2   Broadside Arrays

When one only wishes to realize a high-directivity antenna, the signals are fed to all the patches with the same phase, and the beam points perpendicularly to the air–dielectric interface. The lines of the feed circuit all have the same length.

When all patches are fed with signals of equal amplitude, the power in the side lobes decreases with the angle from the main beam. The first side lobes are then the most important ones and specify the side-lobe ratio. It is also possible to set all the side lobes at the same level by tapering the signal amplitude as a function of the distance from the center of the array. The side lobe ratio is then increased without changing the array size. Radiation patterns for homogeneous and tapered distributions are shown in Figure 11.11. A similar result is sometimes obtained by removing some of the outside patches (thinned array), in which case the resulting array is no longer periodic. Since the feed circuit is time invariant, interactions between the elements can be accounted for and compensated.

11.5.3   Shaped Beam

Particular illumination patterns can be achieved by adjusting both the phases and the amplitudes of the signals fed to the different patches. In the example in Figure 11.12, a particular requirement for base stations of cellular mobile communication systems had to be satisfied [Sanford et al. 1991]. The main beam is pointed slightly below the horizon, tapering sharply above but gently below (cosecant square pattern). The signal amplitude at ground level remains almost constant, providing an even coverage without holes in the pattern.

11.5.4   Phased Arrays

Radar beams continually scan the horizon and the sky, looking for various kinds of vehicles or obstacles. In conventional radars, scanning is done

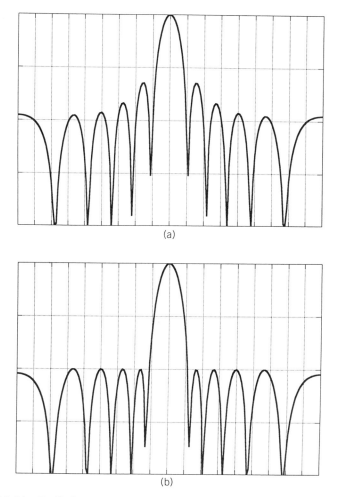

FIGURE 11.11 Radiation patterns for broadside arrays: (a) with homogeneous distribution, (b) with tapered feed.

mechanically by rotating and tilting the antenna. This function can be carried out much faster electronically with a fixed antenna by using a phased array [Hansen 1966]. The direction of the main beam is shifted by changing the phases of the signals fed to the radiating elements. Phased arrays on satellites can similarly direct beams to locations on the ground without modifying the antenna position.

In the array of Figure 11.10, the element (m, n) is located at

$$x_m = x_0 + ma, \qquad y_n = y_0 + nb, \qquad (11.1)$$

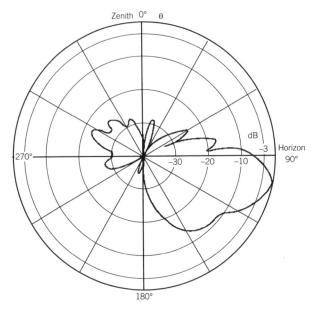

FIGURE 11.12 Shaped beam for base station of cellular mobile communication system. (Courtesy Huber & Suhner Co.)

where a and b are, respectively, the array spacings in the x and y directions, and (x_0, y_0) are the coordinates of the element of reference $(0, 0)$. To point the beam toward the direction θ, ϕ (with respect to the coordinate axes, Figure 11.10), the signal fed to the element (m, n) must be phase-shifted by [Pozar 1989]

$$\Psi_{mn} = -\omega\sqrt{\varepsilon_0\mu_0} \, \sin\theta(\, ma \cos\phi + nb \sin\phi). \qquad (11.2)$$

This value does not take into account the interactions between patches. It is actually correct for infinite arrays, where interactions affect all the elements of the array in the same manner.

11.5.5   The Infinite Phased Array

The properties of a patch in the center of a very large array can approximately be determined by assuming that the array dimensions are infinite. All the elements of the array are then equivalent, and periodicity relationships can be used. In the integral equation techniques (Section 10.6), the Poisson sum formula replaces the Sommerfeld integral by infinite summations [Pozar and Schaubert 1984]. The input impedance and the radiation pattern of a patch within an infinite array are then determined [Pozar 1989]. The computations are not more involved than those for a single patch.

The study of infinite arrays shows that at some scan angles the reflection factor of the elements becomes equal to unity. The elements actually reflect all the power fed to them and do not radiate any of it. In microstrip arrays this condition, called scan blindness, is associated with the presence of surface waves on the antenna, which are strongly excited at synchronism with Floquet modes of the periodic structure [Pozar and Schaubert 1984]. The reflection is total in infinite arrays only, whereas in large arrays it may become significant enough to severely downgrade performance.

11.5.6   The Finite Phased Array

Small phased arrays were analyzed by techniques developed for the study of single patches, considering the array as one multiply connected antenna. This approach is, however, limited by the capabilities of the computer.

The performance of large arrays can be determined approximately by using the parameters derived for infinite-array elements in the summation process. However, the properties of patches close to the array's edges are not the same as those of patches closer to the center, and the accuracy provided by this approach decreases when the size of the array decreases.

A novel method makes use of a convolution technique introduced by Ishimaru et al. [1985] to derive the characteristics of a finite array from those of an infinite one [Skrivervik and Mosig 1992]. This approach is numerically quite efficient and was shown to remain accurate for small arrays and even for single patches.

11.6   FINITE-SIZE ANTENNAS

The integral equation formulations (Section 10.6) are developed under the assumption that the ground plane and the substrate have infinite transverse dimensions. In practice, however, the size of the antenna is finite, and diffraction on the edges produces radiation toward the back of the antenna (Section 1.5.2). This information is required when patches are used as freestanding structures and where the back-to-front ratio must be minimized for interference reasons. The diffraction from the edges of the ground plane may also cause scalloping of the main beam and slightly affect the directivity.

The method of moments [Shafai and Kishk 1989] was applied to determine the current distribution on the conductors and the equivalent polarization current in the dielectric slab. More recently, Bokhari et al. [1992] introduced a hybrid approach, combining the mixed potential integral equation and the weak form of the conjugate gradient–fast Fourier transform (WCG-FFT) [Zwamborn and Van Den Berg 1991]. The current on the patch is calculated for the infinite structure and used to evaluate the equivalent polarization currents in the substrate. The current distribution in the finite patch is then determined by the WCG-FFT method. Since matrix inversion is not required in this last step, the method can be used to analyze small to moderate-size ground planes.

11.7 GENERAL REMARK

Printed antennas were often hailed as an inherently low-cost technology and, thus, are potentially suitable for low-cost consumer applications such as direct broadcast satellites (DBS). However, current designs generally remain expensive due to the cost of substrate materials. Because microstrip antennas are much larger than microstrip circuits, the usual substrate materials become quite expensive and may not be available in large enough sizes. More reasonably priced printed circuit boards (epoxy fiberglass) generally exhibit large losses at microwaves. Polypropylene substrates (Section 13.1.7) combine low loss and low cost but melt at a low temperature [Demeure 1986]. Foam materials possess very low permittivities and look quite promising (Section 13.1.8), but precautions must be taken to keep moisture out.

CHAPTER TWELVE

Computer-Aided Design

12.1 DEFINITIONS

12.1.1 The Need for CAD

In traditional electronics and microwaves, components are manufactured individually, tested, and checked. Defective elements can be picked out and rejected, so only the ones that meet the specifications are kept when assembling a system. On the other hand, integrated circuit techniques produce complete assemblies, in which it is difficult, or even impossible, to characterize individual parts. Should a single element prove defective, the whole assembly may not function correctly and must be rejected. Even then, it may be difficult to pinpoint the offending element.

One must determine how the operation of an assembly made by many components connected together is affected by variations in their parameters, which result from fabrication tolerances and environmental changes. This study is rather complex and requires repetitive operations which are very tedious to carry out by hand. The process is, however, well suited for computers; hence, the development of computer-aided design (CAD).

CAD software packages were first developed for low-frequency electronics. High-frequency and microwave designs are inherently more difficult to study, so the software choice remained limited and expensive for a long time. Furthermore, these CAD packages required full-size computers. The more traditional "cut-and-try" empirical methods are still often used to realize circuits, despite their obvious drawbacks and inadequacies.

The situation changed radically with the advent of personal computers (PCs), which provide extensive and versatile computing power at a relatively low cost. Increasingly sophisticated microwave CAD packages became available. Faced with a wide variety of choice, the designer must now determine which software meets the requirements.

To be effective, CAD techniques require accurate mathematical descriptions, which should also be simple [Hofmann 1984]. The study of microstrip circuits presents a formidable challenge because it is difficult to define models that are simultaneously easy to use and accurate. Most models use equivalent circuits, while some are based on electromagnetic field analysis, in which case precautions must be taken to keep computation requirements reasonable [Gardiol 1992].

12.1.2 Analysis

Theoretical analysis determines the electrical response of a circuit that is completely defined and excited by a specified signal. Analysis is a "direct problem," in which one simulates the operation of a completely known physical structure to find its electrical characteristics. The response is generally unique. The simulation process is analogous to an actual measurement. Theoretical results can be compared with experimental data, allowing one to ascertain the validity of the techniques used.

12.1.3 Sensitivity Analysis

The effect of a change in any parameter is determined by introducing the new value in the computer program and comparing the output data. The sensitivity of a system to each of its parameters is determined in this manner. Varying some parameters may result in a significant change in the operation of the circuits, while others may vary considerably without significant effect. Tolerances on the parameters can then be specified.

The sensitivity of a system is much more difficult to determine experimentally, since one must physically realize and measure many different circuits. Care is required to ensure that only one parameter is modified between successive circuits.

12.1.4 Synthesis

The circuit designer faces quite a different situation: the sales department of the company provides a set of specifications, generally originating with a potential customer, and requests the development of a system to meet those specifications. Practical constraints further limit the designer's choice: the design must use realizable circuit elements and available devices, the circuit should fit within a specified package or on a chip having a specified size, and its fabrication should not be too expensive. This is an "inverse problem," in which the physical configuration is not known a priori. The designer does not know beforehand whether it is possible to build a circuit meeting the requirements. In some situations, several designs may yield the desired results.

In some situations, the analysis procedure (Section 12.1.2) can be mathematically inverted, yielding relationships for the geometry and material parameters in terms of the required response. This is called *direct synthesis* and is generally limited to the design of elementary components (couplers, Sections 6.5 and 6.6, filters, Sections 7.3 and 7.4) and matching circuits.

In the software package MULTIMATCH (Ampsa, for IBM PC), developed by Abrie [1986], matching circuits are directly synthesized for stable transistor operation, with either lumped components or transmission line sections. This approach provides the user with several synthesized designs to choose from, together with their calculated electrical performances. On the other hand, it does not optimize the circuit.

12.1.5   Optimization

For more complex circuits, the design is realized by indirect synthesis. Analysis software is inserted into an optimization loop (Figure 12.1) and used to calculate responses for successive sets of parameters, which are then compared to the desired response until the design meets the requirements. Two strategies can be used: iterative techniques and stochastic approaches.

In an iterative, or gradient, technique an "approximate" design is first introduced in the analysis program, and the calculated response is compared to the specifications. A circuit parameter is then slightly modified, and the response of the modified circuit is determined. If the new response is better than the previous one, the change is favorable and the parameter considered is again modified until subsequent changes do not bring any further improvement. On the other hand, if the new response is not as good as the previous one, it means that the change was made in the wrong direction and the process must be carried out in the other direction. This operation is repeated

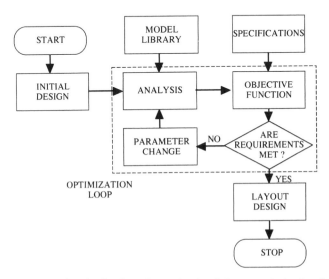

FIGURE 12.1   Synthesis of a microstrip circuit in an optimization loop.

with the other parameters of the circuit. The whole process can then be repeated any number of times until the circuit response reaches a stable value, which may or may not meet the specifications.

In the stochastic or Monte Carlo approach, sets of parameters are randomly generated, and the calculated circuit responses are compared. The "best" response is stored and compared with the specifications.

The specifications generally involve several terms of the scattering matrix (Section 4.3.3) and some other characteristics as well. For an amplifier, one requests that

1. The reflection factor at the input ($|\underline{s}_{11}|$) must be smaller than a specified upper bound.
2. The reflection factor at the output ($|\underline{s}_{22}|$) must be smaller than a specified upper bound.
3. The gain ($|\underline{s}_{21}|$) must be larger than a specified lower bound.
4. The noise figure (LF, Section 8.5) must be smaller than a specified upper bound.

These requirements must be satisfied over the frequency band of the device. One would also like to have a more or less constant ("flat") gain. In addition, the operation must remain stable over the entire frequency band in which the active device (transistor) can amplify. Spurious oscillations may otherwise occur. For power amplifiers, the dynamic range is also specified.

The whole process of optimization can be formulated as the minimization of an objective function (cost or penalty function), which represents the difference between the performance achieved and the desired specifications. Weighting functions are used to specify which characteristics are the most significant.

Several search schemes were developed [Gupta et al. 1981; Rosloniec 1990; Dobrowolski 1991]. Gradient search techniques generally converge toward a minimum of the objective function but cannot determine whether this minimum is the lowest one. Random processes can provide a broader picture but entirely lack the convergence process. These may be used to determine a starting point in the vicinity of the lowest minimum, and gradient search can then be used to pinpoint its location.

12.2   A WORKED-OUT EXAMPLE

12.2.1   Presentation of a Typical Problem

The basic concepts of CAD are easier to understand with a detailed, step-by-step description of an actual design process. The problem most often encountered in practice is that of matching a device over a specified frequency band. The example presented considers a single-stage, low-noise microstrip amplifier for the 5.925- to 6.425-GHz frequency band.

12.2.2 DragonWave

The software selected for this presentation, DragonWave, takes advantage of the user friendliness of the Macintosh and is thus particularly easy to use and well suited for beginners. By simply dragging pictures of the elements over the drawing area on the computer screen and using the mouse to interconnect them, the user automatically creates the schematic of the circuit and the associated node list. Modifications are straightforward with the cut, copy, and paste commands. The DragonWave library contains more than 50 elements, including microstrip, stripline, and coupled lines. Powerful random and gradient optimizers are included, and the program automatically chooses nonnodal or nodal analysis to give the benefits of both. The number of circuit elements, nodes, frequencies, graphs, and reports is only limited by the computer's memory. Graphics and data produced by DragonWave are directly available for insertion into documents. The software is available from Nedrud Data Systems in Las Vegas, Nevada.

12.2.3 Definition of the Circuit

The substrate parameters and the frequency range are first entered into the program (Figure 12.2). The transistor to be used is a NE20283 packaged HEMT from NEC, chosen for its electrical specifications. This transistor's characteristics are stored in the DragonWave library, and the transistor is

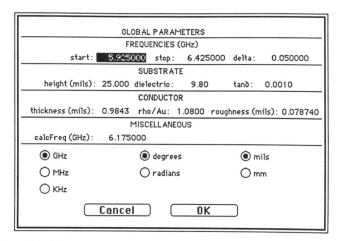

FIGURE 12.2 Definition of substrate parameters and frequency band. (Courtesy B. Nedrud.)

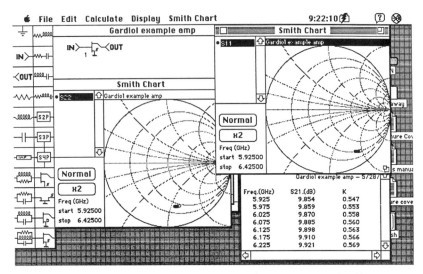

FIGURE 12.3   Transistor parameters. (Courtesy B. Nedrud.)

selected by double-clicking the icon for the grounded-source FET and finding its code in the Trial Lib. The input and output matches are plotted on Smith charts, and a Report window is opened to show the gain and the K-factor (Figure 12.3). This provides the starting point for the matching process.

12.2.4   Matching of the Transistor

The input is matched first, since the output match is less sensitive. Since $\underline{s}_{11}$ is in the third quadrant in the Smith chart, a section of transmission line will move it into the fourth quadrant, where a capacitive stub will bring it toward the center. The output match requires the same elements. Matching is done directly in microstrip, using 50-Ω transmission lines and 65-Ω stubs.

Before connecting the elements, the function AutoCalc in the Calculate menu must be disabled. This function provides automatic calculations while changes are made in the circuit, and breaking the input-to-output connection while AutoCalc is on would trigger an error signal and discard the data. Adding the matching elements to the HEMT then yields the complete circuit (Figure 12.4). After connection, the AutoCalc is reenabled.

The transmission line length and the stub length are then specified by clicking on the respective icons, typing trial values in the window, and pressing the Return key to enter them in the program. A 50° series phase shift and a 75° open stub bring the $\underline{s}_{11}$ pattern close to the center of the Smith chart (Figure 12.5).

The same process is repeated for the output matching circuit, with about 60° phase shifts for both lines. Adding source stability circles to the $\underline{s}_{11}$ chart and load stability circles to the $\underline{s}_{22}$ chart yields the plots of Figure 12.6.

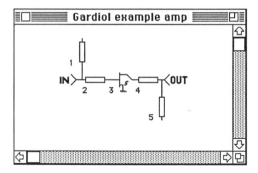

FIGURE 12.4   Transistor with matching elements. (Courtesy B. Nedrud.)

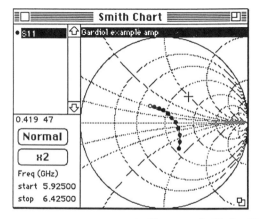

FIGURE 12.5   Input match. (Courtesy B. Nedrud.)

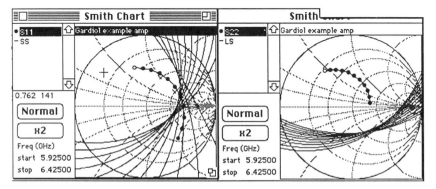

FIGURE 12.6   Input and output match, showing stability circles. (Courtesy B. Nedrud.)

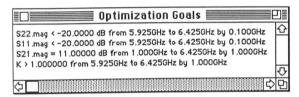

FIGURE 12.7   Optimization goals. (Courtesy B. Nedrud.)

Comparing the input match of Figure 12.6 with the one in Figure 12.5 shows that, unfortunately, the output matching process interacts with the input, moving the plot away from the center of the chart. At this point, it becomes preferable to optimize further steps in the process, since there are many interacting variables.

12.2.5   Optimization Process

Optimization Goals are chosen from the Calculate menu, and Add Goal from the Opt. Goals menu. The four goals entered are shown in Figure 12.7. The parameters to be optimized are then specified (Figure 12.8): line lengths and impedances. It is not recommended to optimize with respect to many parameters because the process may become lengthy. For instance, one might optimize the lengths only and adjust the impedances by hand.

Gradient optimization is done by choosing Gradient in the Calculate menu. The number of iterations is specified, for instance, two or three steps, and the process is repeated until no further improvement is noticed. In the present case, the gain is too frequency dependent and the operating point too near the stability limit, so that the weights on K and S12 are increased and further optimization is performed, yielding Figure 12.9.

At this point, the results obtained are still not good enough, so another matching stub with another transmission line section is added at the input. Before doing that, however, the limits of optimization of the existing transmission lines are set at $\pm 5°$ so that the added matching works with the existing design. The AutoCalc function must again be disabled before adding new elements. Further optimization then yields the data of Figure 12.10. The process is further repeated, and new elements added as required until the specifications are met.

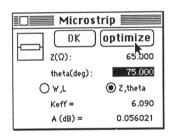

FIGURE 12.8   Optimization parameters. (Courtesy B. Nedrud.)

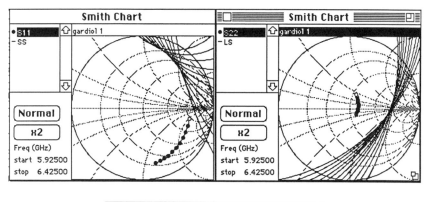

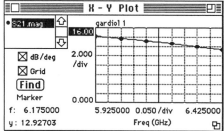

FIGURE 12.9   Characteristics of amplifier after some optimization. (Courtesy B. Nedrud.)

Freq.(GHz)	K	S11.(dB)	S22.(dB)	S21.(dB)
5.925	0.652	-2.754	-15.568	12.017
5.975	0.664	-2.984	-14.624	12.013
6.025	0.675	-3.236	-13.630	12.015
6.075	0.684	-3.510	-12.635	12.022
6.125	0.694	-3.823	-11.689	12.021
6.175	0.705	-4.175	-10.801	12.009
6.225	0.719	-4.565	-9.980	11.980
6.275	0.733	-4.986	-9.228	11.930
6.325	0.750	-5.425	-8.550	11.852
6.375	0.769	-5.854	-7.949	11.739
6.425	0.790	-6.233	-7.426	11.585

Report — Gardiol 3 — 5/29/92 9:17:48 AM

FIGURE 12.10   Characteristics of amplifier after further optimization. (Courtesy B. Nedrud.)

12.2.6   Other Features

The program also provides the noise figure, and to ensure that there are no stability problems outside of the operating frequency band it is recommended to determine the response over a very wide band. Finally, secondary effects can be taken into account: bias lines, blocking capacitors, discontinuities, and so on.

12.2.7   Remark

The example shows that, when using the DragonWave program, the design and optimization processes are interactive. The designer has a possibility of "seeing" what he or she is doing during the complete process. A parameter change produces changes that are immediately apparent on the computer screen, either on Smith charts or on X-Y plots. Visual interaction between the program and the user is a most significant feature for the understanding and general check of the design procedure. Numerical data, which are more precise but more difficult to interpret, can be obtained in Reports.

12.3   SOME CAD PACKAGES

Many software packages have become available in recent years, providing features like those illustrated in the previous example, while some of them are even more elaborate. Surveys of available CAD software occasionally appear in the technical literature, providing some comparisons [Besser 1984; Pitzalis 1989]. A brief description of the better-known packages follows. The section heading is the name of the company, while the names of the software packages are listed in the text.

12.3.1   Compact Software

Supercompact is a general analysis and optimization design tool for microwaves and rf (for VAX, IBM, IBM-PC, Apollo, PC, HP 9000). Four-port circuit analysis and optimization can be achieved, as well as transistor impedance modeling (transistor library) and matching network synthesis. Single, coupled, and interdigited microstrip designs can be created. Microstrip and other planar lines can be specified in terms of their electrical and physical dimensions; approximate expressions are included, with accuracies of 1% or better. The effects of dispersion, radiation, discontinuities, multilayers, metallization, surface roughness, and dielectric and conductor losses are taken into account. An FFT time-domain option is available. Circuits are optimized by the random perturbation and gradient techniques, while their parameter sensitivity is determined with a Monte Carlo algorithm.

The electromagnetic simulator Explorer provides a rigorous characterization of complex geometries, handling planar multilevel structures [Hill and Kottapalli 1992].

12.3.2   EEsof

Touchstone (for IBM-PC and HP series 200) is a fast general-purpose rf and microwave circuit design, analysis, and optimization program specifically developed for PCs. More than 80 elements are contained in its catalog, including microstrip, stripline, waveguide and coplanar lines and discontinuities, lumped elements, and electronic device models. The user can also define elements with their scattering matrix. Possibilities for adjustments are offered by two optimization algorithms (random perturbation and gradient method) for up to 15 variables. The companion program Monte Carlo determines the sensitivity of the circuit. EEsof also provides device synthesis programs compatible with Touchstone: waveguide, microstrip, and stripline band-pass, low-pass, and band-stop filters, and coaxial low-pass and band-stop filters. The program E-Syn synthesizes matching circuits. Microwave SPICE provides nonlinear circuit analysis and synthesis. The program Anacat is geared toward measurements, in particular with vector network analyzers, and provides embedding and deembedding functions. Libra combines linear analysis in the frequency domain with time-domain analysis of nonlinear elements. It is ideal for the analysis of digital circuits driven by fast, periodic clock pulses, where the effects of transmission line mismatch and delay line effects are critical. Powerful hierarchical circuit and system design can be done by using Academy as an efficient multitasking controller for most of EEsof packages.

12.3.3   Hewlett Packard

The Hewlett-Packard (HP) microwave linear simulator (MLS) simulates and optimizes linear circuits in the frequency domain. It was developed for integration into the microwave design system (MDS), a unified system of professional computer-aided engineering (CAE) tools. Designers use the design capture system to assemble a circuit schematic, which forms the input to the MLS, which then simulates the circuit and displays the results with a built-in presentations package. Circuits having up to nine ports and a virtually unlimited number of subcircuits can be simulated, with up to 500 nodes for each. Designers can add user-defined models by directly programming the scattering matrix parameters of a "black box" element. The MLS can adjust the circuit's parameters automatically to improve circuit performance, alternately using random and gradient optimization. The schematic is then updated with the improved values.

The MLS implements a number of advances in software technology, combining accurate test models with advanced sparse matrix techniques in a professional package unbounded by the constraints of the PC environment. The complete MDS system runs under the HP-UX operating system on HP 9000 series 300 workstations.

12.3.4 Jansen Microwave

Jansen [1985] developed the software package LINMIC (for HP 9000 series 500 and 300, Microvax II) which combines a field analysis, based on an enhanced spectral-domain technique, with equivalent circuits for simple and coupled lines, discontinuities, T-junctions, capacitances, resistors, transistors, and other lumped elements [Jansen 1986]. The LINMIC package can describe structures for which no analytical models have been developed: interdigited capacitors, multiturn square inductors, coupled meander lines. Up to four dielectric layers can be considered, including passivation and metallization layers, ohmic and dielectric losses, coupling, and higher-order mode effects. Up to 40 connections to the outside can be considered, and the structure can be divided into 20 sublattices. An interactive optimization procedure is provided, based on a conjugate direction algorithm, in terms of reflections, gain, insertion losses, ripple, noise, and stability, as defined by 60 or more parameters. The sensitivity with respect to those parameters is analyzed. Very high accuracy, a wide validity range and a high computation speed drastically expand the range of applications. It is highly portable since 98% of its code is written in FORTRAN 77.

12.3.5 Sonnet Software

The program Em carries out a complete (full-wave) three-dimensional (3D) electromagnetic analysis based on the method of moments (Section 10.6.9). It rigorously includes stray coupling, discontinuities, interactions between discontinuities, higher-order modes, package resonances, and dispersion, dielectric, metal, and radiation loss. The program analyzes any multilayer predominantly planar circuit, including short via holes (Section 13.6.7), wire bonds, and air bridges, yielding the scattering parameters or the current distribution as a function of frequency. It can evaluate entire interstage matching networks of high complexity and is generally used to validate the portions of a design where strong circuit interactions and stray coupling are present.

Em runs under UNIX on Sun, HP, Apollo, IBM Risc 6000, DEC, and Cray. The X-Window System is required, with 8 Mbytes or more of memory. The circuit geometry is specified by xgeom, a mouse-based graphical user interface. Output data is provided in Touchstone or Supercompact format, and SPICE files can also be generated. The distribution of the surface current can be represented in color with the emvu software.

12.3.6   ArguMens Software

ArguMens, based in Duisburg, Germany, proposes a line of software packages named after sea creatures. Octopus is an advanced microwave circuit design and noise analysis program for the analysis of circuits having from one to six ports. Special routines were developed for device modeling and circuit optimization using the Evolution strategy and modified Powell Fletcher routines. Tolerance analysis uses Monte Carlo statistics and two kinds of worst-case approaches. An extensive element library includes modern and advanced models considering the requirements of MMICs. Lamprey is a dedicated software package for the optimization with respect to measured scattering parameters. Lobster is a program for microstrip and other rf and microwave antennas. The program Quick-Wave analyzes arbitrarily shaped two-dimensional microwave circuits.

12.3.7   Caro-Line Habigand

The software package (technology oriented analog simulation tools) TOAST is devoted to the design of analog microwave circuits. It contains a linear simulator including the analyzer, the computation of sensitivity, several optimization schemes, statistical analysis, and a library of models, together with a nonlinear simulator based on the spectral balance approach.

12.3.8   Thom'6

ESOPE (for Vax, IBM, Apollo) is an interactive software for the analysis and optimization of linear microwave circuits. Most components (lumped RLC, transmission lines, waveguides, coupled lines, microstrip) are implemented. Active circuits are described by their scattering parameters. Optimization can be made with one or several objectives, three algorithms are available (min-max, least squares, fixed tolerance), the sensitivity and worst case are determined, and results are represented graphically. The program is in FORTRAN 77. Drafting of the circuit's outline is carried out by the software Hyper'6-D.

12.3.9   Spefco Software

CiAO (for IBM PC-XT or AT) is a general analysis and optimization program for circuits composed of RLC lumped elements, controlled sources, gyrators, lossy or lossless transmission lines, one- and two-ports described by their scattering matrix. The companion program, Design, synthesizes wideband matching circuits for linear-gain amplifiers.

12.3.10 Made-It-Associates

The software measurement and microwave analysis (MAMA), for HP 9836 or HP 9000 series 300, analyzes and designs quarter-wave transformers, Lange couplers, directional couplers, microstrip lines and discontinuities, hybrid circles, and power dividers. It synthesizes filters and rectangular printed antennas. It can be used for deembedding components and it interfaces with Microcompact.

12.3.11 Radar Systems Technology

The package ANALOP (for IBM-PC and CP/M-80) analyzes and optimizes, on 15 variables at 15 frequencies, two-ports having up to 200 elements: lumped *RLC* components, series and parallel resonators, transmission lines, transformers, impedance inverters, microstrip transmission lines and discontinuities, rectangular waveguides (dominant mode), controlled sources, filters, and couplers.

12.3.12 Webb Laboratories

The program TRANSCAD (for IBM PC-XT/AT) is devoted to the study of transitions between transmission lines (coaxial, bifilar, planar structures) and waveguides.

12.3.13 Optotek

The monolithic and microwave integrated circuit analysis and design (MMICAD) program performs analysis and optimization of active and passive electronic circuits. It is suited to all types of circuits, from low frequencies to microwaves. It includes one of the fastest linear simulators, complete nodal analysis including nodal noise capability, three different optimizers, device scaling, and network deembedding functions. Through its use of advanced dynamic memory allocation techniques, complex simulations can be performed rapidly on any IBM or compatible PC. In-depth help information and extensive error checking are provided throughout the program. MMICAD is distributed by Optotek Ltd., Kanata, Ontario.

12.4 COMPUTER-AIDED LAYOUT

When a design has been analyzed with the most accurate models available and optimized by a sophisticated CAD package, one of the most critical parts of the procedure remains to be carried out—its physical realization. Before moving on to the photolithographic process, the pattern of the upper conductor must be drawn and transferred to make the mask. Nowadays, coordinatographs and photoplotters can generate very accurate masks, but these machines are expensive and beyond the reach of small research laboratories or academic institutions. The layout and cutting of masks can also be done at a much lower cost on a standard plotter.

12.4.1 Autoart

The Autoart program converts microwave circuit models described by their physical dimensions (single and coupled transmission lines, circular radial stubs, tapered lines, Lange couplers, circuit discontinuities) into artwork. Circuit data provided by the Supercompact package is used to prepare the conductor pattern, and the geometric data are then transferred to a coordinatograph or a photoplotter, with which Autoart interfaces directly [Laverghetta 1991].

12.4.2 Micad

In a similar manner, the data provided by the general-purpose program Touchstone can be transferred through the Micad software to generate a mask on a coordinatograph or a photoplotter [March 1984].

12.4.3 Calma

Calma is an interactive graphic system that can be used to lay out microwave circuits by specifying the x, y coordinates of the main points from a drawing. The layout is provided in the form of a coordinate printout and a pen plot of the circuit. The complete circuit is stored on tape that can be fed to a plotter to make artwork. The system consists of a terminal with two displays (one showing the layout and the other the numerical data) and a digitization tablet for entering the data points [Laverghetta 1991].

12.4.4 Shark

The software package Shark is a link between the design program Octopus (ArguMens, Section 12.3.6) and output devices used to produce masks. The geometric data delivered from Octopus is transferred to a set of masks with up to 13 levels. The user may redesign the layout or add user-defined elements. The program checks the geometry of the connections.

12.4.5 Micros

While previous layout softwares are additions to analysis and synthesis packages, the software Micros was developed specifically for the layout of microstrip circuits with HP 9000 series 200 and 300 workstations [Zürcher 1985; Zürcher and Gardiol 1989]. The complete program also provides the operator with the necessary information. The software is distributed by High Tech Tournesol (Lausanne, Switzerland).

The operator specifies the layout characteristics: substrate permittivity and thickness, frequency, characteristic impedance of transmission lines, correction of undercutting (Section 13.3.8). The circuit's outer dimensions are also entered. The computer determines the line widths required to build components. A menu presents the available components (Figure 12.11) that can be displayed on the computer screen, rotated, and positioned. The designer may also add user-defined elements.

The operator selects the components required for the design and places them on the layout with the proper orientation. The ports that must be connected together are indicated with a runner on the screen. Compensated mitered bends are used for the connections [Douville and James 1978].

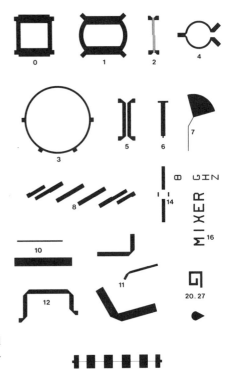

FIGURE 12.11 Components realized with Micros. (Courtesy High Tech Tournesol.)

The completed circuit may then be drawn on paper, where the operator checks the respective positions of the components, the spacing between connecting lines, whether lumped elements can be inserted, and so on. When the design is found to be satisfactory, the mask is cut on a Rubylith sheet placed on a standard plotter equipped with a specially designed cutting tool.

12.5   SELECTION OF A CAD PACKAGE

Many CAD packages are presently available, as evidenced from the sampling presented in Sections 12.3 and 12.4. However, there are still many more, at various levels of price and sophistication [Pitzalis 1989].

12.5.1   A Wide Variety of Software

There is certainly no shortage of CAD software packages for microstrip design, some of which are broad-band general-purpose analysis–synthesis packages and others are specifically devoted to the design of a particular class of components. Many of the programs apparently fulfill similar tasks, making the selection difficult. The simple packages are based on equivalent circuit descriptions, which neglect radiation and interactions and can generally run on PCs. The highly sophisticated packages include electromagnetic simulation and nonlinear analysis in the time domain, but they require powerful mainframe computers. The designer should first determine the real requirements and then carry out the selection accordingly.

12.5.2   Criteria for Selection

Ideally, one should evaluate several packages before making a selection. If possible, a sample circuit should be designed and their respective performances compared.

1. *Cost and Availability*   Is the information provided by the supplier accurate? Are there hidden costs? What services are included in the basic package?
2. *Validity of Results Obtained*   Do the calculated data agree with the values measured (after the realization of the circuit)? The accuracy provided by the simulation process is very important, particularly when optimization is involved. The use of crude models in early CAD packages discouraged many potential users for a long time. Users noted that optimization algorithms did not always perform as claimed by the vendors and did not always converge to the best possible solution.

3. *Ease of Operation*   Is the program user friendly? Are instruction manuals easy to understand and properly documented? Is it possible to closely follow and easily control the optimization process? Are the data presented graphically?

4. *Computer Time and Memory Requirements*   Is a PC sufficient to carry out the computations, or is a more expensive workstation or a mainframe computer necessary? Do you have easy access to such a computer?

5. *Compatibility*   Can input and output data files be easily transferred from or to other design tools (measurement, layout). Transcription of files is time consuming and likely to produce errors.

6. *Transportability*   Will the software still work when the operating system of the computer is upgraded or when a new computer is installed? Some software packages can only be installed once.

7. *Technical Support, Updates, Debugging, and Other Maintenance*   Service is a key nontechnical parameter in actual operation.

The *International Journal of Microwave/Millimeter Wave Computer-Aided Engineering* presents, as a regular feature, the resolution of sample CAD problems with different software. This provides an interesting base for comparison. The selection is, however, hampered by licensing "agreements" imposed by some software producers, which require confidential treatment of information provided and specifically prohibit the comparison (public or private) of their wares and competitor's products!

12.5.3   Remarks and Comments

Many models used to characterize microstrip lines and discontinuities are remarkably accurate nowadays. One must remember that they are derived under some simplifying assumptions; that is, lines are generally assumed to be infinite and uniform, far enough from walls or substrate edges to neglect interactions. It is still difficult to account for covers and walls close to the circuit or for coupling between discontinuities. Such situations often occur in monolithic microwave integrated circuits, where components are located very close to one another to save real estate. Only the recent sophisticated electromagnetic simulations packages can take such effects into account (Sections 12.3.1 and 12.3.5).

With proper care, a fairly good correspondence can be obtained between theory and experiment for passive components, but this is not always true when active devices are involved: measured amplifier gains are often some dBs below predicted values, even when line losses are taken into account.

Also, computation accuracy is not everything. Material properties are often not known accurately enough. It is not uncommon for substrate thickness to be specified within 5% or 10% (!!), and the accuracy on the permittivity of composite materials may be of the same order. It is recommended to measure the substrate before starting with the design of frequency-dependent circuits [Zürcher et al. 1986].

One must never forget that CAD is a tool that significantly boosts the capabilities of a skilled operator, but which is useless and wasteful in the wrong hands. The Ga-Ga principle (garbage in, garbage out) applies: for instance, when the substrate properties are not accurately known, one should not expect to obtain the proper frequency dependence. An expert designer will detect potential problem areas, such as lines too close to one another or too close to a wall or edge. On the other hand, it would be extremely difficult to include all possible interactions in a computer program. In microstrip design, as in many other technological fields, the computer will not completely replace the skilled designer.

CHAPTER THIRTEEN

Fabrication Techniques

The microstrip structure is defined in Section 1.1.1. It consists of a thin plate of low-loss insulating material, the *substrate*, completely covered with metal on one side and partly on the other, where the circuit or antenna patterns are printed (Figure 1.1).

13.1 THE SUBSTRATE

13.1.1 Electrical Properties

The substrate is an integral part of transmission lines, circuit elements, and deposited components, which determine the electrical characteristics of the circuit or antenna. The main electrical parameter, which specifies the size and the nature of the circuit, is the relative permittivity ε_r of the dielectric. Together with the substrate thickness h, the permittivity defines the basic operation of the structure (Section 1.3.6). A thin substrate of high permittivity is required for most lines and circuits, whereas a thick substrate of low permittivity ensures proper radiation conditions in an antenna. The value of ε_r determines the size: when other parameters are kept equal, the size of a circuit is roughly proportional to $\sqrt{2} / \sqrt{1 + \varepsilon_r}$. Size reduction is desirable for low-frequency operation (typically below 1 GHz). A good substrate material should have a uniform permittivity ε_r and thickness h over the whole structure, as well as for different batches of a given material, to ensure circuit reproducibility.

The dielectric losses of the substrate should be small (ideally, $\tan \delta < 0.001$) in order to ensure high performance of circuits and acceptable quality factors for resonators, filters, and radiating elements.

13.1.2 Mechanical Properties

Since the substrate is also the mechanical support of the structure, it must satisfy mechanical requirements:

Mechanical strength (characterized by the breaking point), which defines impact and vibration resistance

Shape stability, especially for encasing

Expansion factor, which should be small and as close as possible to that of the metal used for the conductors and the enclosure

Long-term stability in the presence of hostile environmental conditions (moisture, temperature cycling)

13.1.3 Physicochemical Properties

The following physicochemical parameters are significant:

Mechanical stability up to high temperatures (soldering, deposition of components in the thick-film technique)

Resistance to chemicals, particularly during the photolithographic process (Figure 1.2)

Surface flatness (bending tends to render the encasing procedure difficult)

Smooth surface, to reduce losses and ensure a good adhesion of conductors

Easy machining, for cutting and drilling holes

Low water absorption, since water exhibits a large permittivity and high losses

13.1.4 Production Requirements

In addition, substrates must meet some very practical requirements if they are to be used successfully in large-scale production:

Low cost

Guaranteed availability of the material

Availability of adequate sizes

Nonhazardous machining

Nonpolluting material during the fabrication process, the actual utilization of the device, and its final disposal

13.1.5 Substrate Materials

Considering the requirements in Sections 13.1.1–13.1.4, it is fairly obvious that no substrate could simultaneously meet all of them. For every applica-

TABLE 13.1   Properties of Some Microstrip Substrates

Material	Relative Permittivity (ε_r) at 10 GHz	Loss Factor $(\tan\delta)$ at 10 GHz	Material	Relative Permittivity (ε_r) at 10 GHz	Loss Factor $(\tan\delta)$ at 10 GHz
Alumina 99.5% Al_2O_3	9.5–10	0.0003	RT/Duroid 5870	2.33 ± 0.02	0.0012
Alumina 96% Al_2O_3	8.9	0.0006	RT/Duroid 5880	2.2	0.0009
Alumina 85% Al_2O_3	8.0	0.0015	RT/Duroid 6002	2.94	0.0012
Beryllia BeO	6.4	0.0003	RT/Duroid 6006	6.0 ± 0.15	0.0019
			RT/Duroid 6010.5	10.5 ± 0.25	0.0024
$(Zr, Sn)TiO_4$	38	< 0.0001	Ultralam 2000	2.5 ± 0.05	0.0022
BaO—PbO—					
Nd_2O_3—TiO_4	88	< 0.0001			
DI-MIC CF	21.6 ± 0.6	0.0003	TMM-3	3.25	0.0016
DI-MIC CB	29.0 ± 0.7	0.0004	TMM-4	4.5	0.0017
DI-MIC CD	37.0 ± 1	0.0004	TMM-6	6.5	0.0018
DI-MIC CG	67.5 ± 2	0.0008	TMM-10	9.8	0.0017
DI-MIC NR	152.0 ± 5	0.0010	TMM-13	12.85	0.0019
Trans-Tech D-MAT	8.9–14	< 0.0002			
Trans-Tech D-450	4.5	< 0.0004	Arlon DiClad 527	2.5 ± 0.04	0.0019
Trans-Tech S-145	10.0	< 0.0002	Arlon DiClad 870	2.33 ± 0.04	0.0012
Trans-Tech S8400	10.5	< 0.0001	Arlon DiClad 880	2.20 ± 0.04	0.0009
Trans-Tech S8500	38.0	< 0.0001	Arlon DiClad 810	10.5 ± 0.25	0.0015
Trans-Tech S8600	80.0	< 0.0003	Arlon Epsilam-10	10.2 ± 0.25	0.0020
			Arlon CuClad 250	2.4–2.6	0.0018
Polypropylene	2.18 ± 0.05	0.0003	Arlon CuClad 233	2.33 ± 0.02	0.0014
			Arlon CuClad 217	2.17 ± 0.02	0.0008
Silicon Si (10^3 Ω-m)	11.9	0.0004	Arlon IsoClad 917	2.17 ± 0.02	0.0011
GaAs (> 10^3 Ω-m)	13.0	0.0006	Arlon IsoClad 933	2.33 ± 0.02	0.0014
Ferrite	9.0–16.0	≈ 0.0010	Epoxy FR4 GE313	4.4	≈ 0.0100

tion, one must carefully consider the different requirements and select the material that provides the best compromise. Substrates can be grouped in five main categories: ceramic, synthetic, composite, semiconductor, and ferrimagnetic. The relevant data on some materials are given in Table 13.1.

13.1.6   Ceramic Materials

Alumina (Al_2O_3) is the most commonly used ceramic substrate. It is characterized by a good surface quality, very low losses, and very little dispersion between batches. However, it is slightly anisotropic. Common thicknesses are 0.254, 0.635, and 1.27 mm, while dimensions are generally stated in inches (1 in.× 1 in. to 4 in.× 4 in.). Alumina is very hard and brittle (hence, quite difficult to machine), and its permittivity varies with porosity. Since adhesion of copper and gold to alumina is poor, an intermediate layer of chromium (or of some other lossy conductor) is required.

Sapphire is the monocrystalline form of alumina. It is used in very particular applications at very high frequencies, when a very smooth surface is required. Sapphire exhibits crystalline anisotropy.

Other comparable ceramic materials available in substrate form are MgAl titanate, with $\varepsilon_r = 8.9–14$, and magnesia ($\varepsilon_r = 10$).

Beryllia (BeO), with a lower permittivity than alumina, presents an extremely large thermal conductivity, which makes it a particularly useful material for high-power applications (removal of heat produced by semiconductors). Unfortunately, BeO powder is highly toxic, so precautions must be taken while machining it.

Quartz (SiO_2) is also a brittle material ($\varepsilon_r = 3.78$) with a relatively low heat dissipation factor. It is sometimes used at millimeter waves because its low permittivity may help to keep reasonable dimensions. Being a mechanically weak material, quartz could be replaced by glass ceramics such as cordierite ($\varepsilon_r = 4.5–6$) [Howard 1992].

High-permittivity materials exhibit relative permittivities of $\varepsilon_r = 38$ and $\varepsilon_r = 88$ [Tamura et al. 1988]. The use of these materials leads to size reductions, which may be of interest at low microwave frequencies (1 GHz or less). Ceramics with $\varepsilon_r = 21.6–152$ are also available in substrate form (DI-MIC, Dielectric Laboratories Inc.).

13.1.7 Synthetic Materials

Pure synthetic materials like polytetrafluoroethylene (PTFE Teflon) or polyolefin have a low permittivity (ε_r between 2 and 3) and rather poor mechanical properties: distortion, temperature dependence.

Polypropylene has $\varepsilon_r = 2.18$ and low dielectric losses, with $\tan \delta = 0.0003$ at 10 GHz. It is a low-cost substrate material, available in large sizes, and thus interesting for printed antenna feeds [Demeure 1986]. However, it has a low melting point (165° C), so soldering is only done with low-temperature solders ($\approx 125°$ C).

13.1.8 Composite Materials

Adding fiberglass or ceramic fillers can significantly increase the mechanical stability of synthetic materials. The permittivity may also be significantly modified, but losses become larger. Most circuits are presently realized on woven glass-fiber PTFE, microfiber PTFE, or ceramic-loaded PTFE [Laverghetta 1991].

Fiberglass-reinforced PTFEs (such as RT-Duroid 5870) are stable substrates, with a permittivity slightly larger than the original plastic ($\varepsilon_r = 2.17–2.6$). They present some anisotropy, especially for woven fiber materials.

Ceramic-loaded plastics (such as RT-Duroid 6006 and 6010.5) exhibit larger permittivity values, which may approach that of alumina, with favorable mechanical properties. Machining and drilling is easy, so they can replace alumina during prototype development, even though losses are slightly larger. The permittivity of some recently developed materials is temperature compensated.

Thermoset microwave materials (TMM) are based on a novel polymer, which in its finished form is a very highly cross-linked hydrocarbon. They are electrically and mechanically stable, their permittivity is almost independent of temperature, and their coefficient of thermal expansion is matched to that of copper. They are available with relative permittivities of 3.25, 4.5, 6.5, 9.8, and 12.85 [Traut 1989b].

Epoxy–fiberglass boards used for electronic printed circuits can be used at the low end of the microwave band but present rather large dielectric losses ($\tan \delta = 0.01$–0.02 at 10 GHz).

For antenna operation, very low permittivities are desired. This can be achieved with foam or honeycomb materials. For instance, polymethacryl-amid foam (e.g., Rohacell 51) has $\varepsilon_r = 1.07$ (i.e., only slightly larger than air) and a loss factor of $\tan \delta = 8 \times 10^{-4}$. The material is fairly rigid but does not present a smooth surface on which metal could be deposited. The conductors are then realized on a thin dielectric foil (Mylar or epoxy) which is then attached to the foam. Foam material is very porous, since it consists mostly of minuscule air pockets, so precautions must be taken to prevent moisture from seeping into the small cavities.

Some synthetic and composite substrates are available in large sizes, which is very convenient in practice. A slight anisotropy can be tolerated in most usual applications.

Some commercially available materials with their properties are given in Table 13.1.

13.1.9   Semiconductor Materials

Metallic strips deposited on semiconductors (Si, GaAs) are also microstrips. Substrate losses are in this case larger than for dielectrics, even though relatively low losses can be achieved in extremely pure semiconductors (low doping levels). Semiconductors are very significant substrate materials to realize complete MMICs that combine conducting strips with active solid-state devices and antennas. Complex subassemblies such as converters, medium-frequency amplifiers, and broad-band amplifiers may be realized on a single "chip."

13.1.10   Ferrimagnetic Materials

Ferrite and YIG substrates have relative permittivities in the 9–16 range and generally low dielectric loss. When placed in a biasing magnetic field, they exhibit a gyromagnetic anisotropy that permits the realization of nonrecipro-cal devices like isolators and circulators (Section 6.3.2).

TABLE 13.2   Resistivity and Conductivity of Metals

Metal	Resistivity Ω-m $\times 10^{-8}$ at 20° C	Conductivity S/m $\times 10^{6}$ at 20° C
Silver	1.62	61.73
Copper	1.72	58.13
Gold	2.44	40.98
Chromium	2.60	38.46
Aluminum	2.62	38.16
Brass	3.90	25.64
Nickel	6.90	14.49
Platinum	10.50	9.52
Lead	21.90	4.56

13.2   METALLIZATION

13.2.1   Requirements

Metal layers deposited on the dielectric substrate must exhibit the following characteristics:

Low resistivity (small ohmic losses, Table 13.2)

Sufficient thickness, at least three times the skin depth δ (for instance, $\delta = 2$ μm in copper at 1 GHz)

Good resolution (angles clearly defined)

Resistance to oxidation

Good solderability (the soldering material must not diffuse into the conductor)

Suitability for the different contact-making processes (thermocompression, ultrasonic welding, etc.)

Strong adhesion to the substrate material, even when thermally and mechanically stressed

Stability while aging

13.2.2   Adhesion

On most synthetic and composite substrates, all the good conductors (Ag, Cu, Au, Al) are suitable and adhere well. On the other hand, adhesion is very poor on ceramic substrates such as alumina, and an intermediate layer of low-conductivity Cr, Ta, or Ti must be added. The layer must be quite thin and may be used to realize resistors (Section 13.4.2).

13.2.3 Protective Layer

In many instances, one wishes to cover the upper conductor with a protective layer, most often of gold, which resists oxidation and aging. One commonly encounters up to three layers on ceramic substrates: chromium for adhesion, copper for conduction, and a gold flash for protection. Only one or two layers (Cu and Au) are used on synthetic substrates.

13.2.4 Remark

The presence of additional layers, required for ceramic substrates, complicates and lengthens the etching process and is clearly undesirable in some applications, particularly during the development of prototypes.

13.3 PHOTOLITHOGRAPHIC PROCESS

The metallization pattern of the upper conductor, on top of the microstrip substrate, is done by means of photography, exposing a layer sensitive to light through a *mask*. The latter is designed and cut, generally at a larger scale (coordinatograph or plotter, Section 12.4), then photographically reduced to the proper size. It can also be made on a photoplotter.

13.3.1 Subtractive and Additive Processes

The conductor pattern is realized either by removing metal from a completely metallized substrate or by partially covering a previously bare substrate.

13.3.2 Deposition of the Photosensitive Layer

A photosensitive varnish (photoresist) is deposited on the structure, and several processes may be used to do this:

Dipping The whole structure is dipped into the photoresist and pulled out at a constant speed. The layer thickness depends on the withdrawal speed and on the viscosity of the photoresist. This process yields rather thick layers with a bulge on the lower edge.

Spraying The photoresist is sprayed on the structure through a nozzle. It is difficult to obtain a constant thickness in this manner.

Centrifuge The structure rotates rapidly. The photoresist is deposited at the center of rotation, and the centrifugal force sweeps it toward the outside. Thin and uniform layers are obtained in this manner, but the technique can only be used for small substrate sizes.

Heat and Pressure   Some photoresists are deposited by pressing a film and heating it (Riston).

After deposition, the photoresist must be cured at a high temperature so that it becomes tougher and more adhesive.

13.3.3 Exposition through the Mask

The photographic mask is vacuum-pressed upon the structure, with its emulsion next to the photoresist layer. The photoresist is exposed to ultraviolet rays through the mask. For an accurate reproduction of the pattern, the UV radiation must be parallel and the photoresist layer must be thin.

13.3.4 Development

During development of the photoresistive layer in a suitable chemical, either its UV-exposed part (positive photoresist) or its nonexposed one (negative photoresist) is removed. The circuit is then ready for the next step, which is either the deposition or the removal of metal.

13.3.5 Etching

When one starts with an entirely metallized substrate, part of the metal must be removed by etching. The metal is exposed to an acid that dissolves the metal but does not affect the remaining photoresist. The process becomes complex when several metallic layers are involved, since specific etching solutions are required to remove each metal layer. Careful rinsing is necessary between the different baths.

13.3.6 Metal Deposition

In the additive process, one deposits metal on the substrate through the holes in the photoresist. Several processes may be used to do this: vacuum evaporation, sputtering, chemical plating (electroless). The layers obtained are generally thin (2–15 μm).

13.3.7 Additional Steps

The remaining photoresist is removed with a solvent or a concentrated alkaline solution. The metallic layer is sometimes thickened by an additional deposition of electrolytic metal. In some cases a protective layer (for instance gold) is deposited to prevent oxidation.

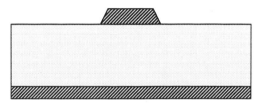

FIGURE 13.1   Conductor cross section resulting from undercutting.

All the steps in the procedure must be carefully separated by rinsing and cleaning operations, sometimes followed by drying in an oven. Drying is particularly critical for some plastic substrates, since some of them tend to absorb water.

13.3.8   Undercutting

The circuit pattern should be as similar as possible to the one of the photographic mask. One must, however, note that the etching process may produce undercutting. This means that the cross section of the upper conductor is not rectangular, but trapezoidal, because the acid etch does not remove metal uniformly (Figure 13.1). The same occurs with metal deposition, where the process does not yield a constant growth. Undercutting is more important when the metal layer is thick: for this reason, thin conductor strips are preferred, as long as the electrical conduction requirements are satisfied. Undercutting is a predictable process that can be taken into account when designing the circuit and cutting the mask.

13.3.9   Thick-Film Techniques

The techniques described realize thin-film circuits, which are used for most microwave circuits. Another approach, the thick-film process, deposits a paste through a mask-covered silk screen. This technique is less accurate and generally not used to realize microwave circuits. On the other hand, it is well suited for components like resistors or capacitors, which can be adjusted individually (Section 13.5.2).

13.4   LUMPED ELEMENTS: DEPOSITED COMPONENTS

13.4.1   Definition

Lumped deposited components are built directly on the substrate and thus become an integral part of the microstrip circuit. Resistors, capacitors, and inductors can be realized by this process. Their realization within the circuit produces fewer discontinuities than the insertion of discrete components, so

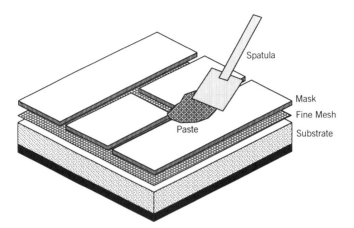

FIGURE 13.2 Silk-screening process.

they are better suited for very high frequency operation. They are also generally smaller than discrete elements (Section 13.5).

13.4.2 Resistors

Resistors can be realized in two ways (Figure 13.3). A resistive paste (palladium oxide mixture) may be deposited by silk screening (thick-film technique, Section 13.3.9) and then cured. This process can only be applied to substrates that withstand high temperatures. Alternatively, the low-conductivity nickel or chromium layers required for adhesion on ceramic substrates can be used to realize resistors in an easy way: one just removes the top metal layers.

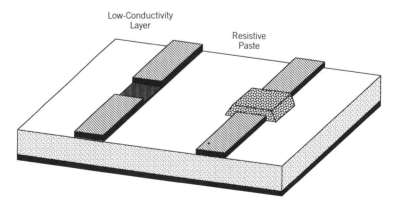

FIGURE 13.3 Deposited resistors.

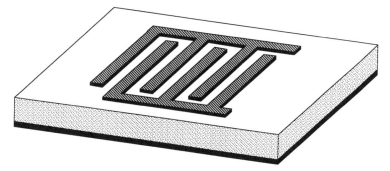

FIGURE 13.4   Deposited capacitor.

13.4.3   Capacitors

Capacitors can be realized by the successive deposition of dielectric and metal layers either by silk screening (thick-film process) or by evaporation (thin-film process). Additionally, a single-layer capacitor can also be made by the photolithographic process by using an interdigital geometry (Figure 13.4).

13.4.4   Inductors

Inductances are also realized by the photolithographic process, in the form of a loop or a spiral (Figure 13.5). In the latter case, the inner connection must be "pulled out" by bonding a thin gold wire or by building a dielectric bridge with a metal strip deposited on top.

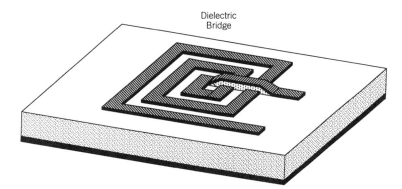

FIGURE 13.5   Deposited inductor.

13.5 LUMPED ELEMENTS: DISCRETE COMPONENTS

Discrete miniature components can also be connected on the microstrip. A number of semiconductor devices, diodes, and transistors, as well as nonreciprocal ferrite devices (isolators and circulators) and dielectric resonators, can also be inserted to realize interconnecting networks and active antenna feeds.

13.5.1 Description

Discrete components are, in general, commercially available and ready for insertion into a circuit. Many common electronic components—active and passive—are presently fabricated at microwave frequencies in a package suitable for insertion within a microstrip circuit. They are generally small compared with a wavelength, and wire connections are extremely short or replaced by metallized connecting surfaces to keep parasitic effects as small as possible. Special device cases reduce the discontinuities as much as possible.

13.5.2 Resistors, Loads, and Attenuators

Resistors are small "blocks" with sides from a few tenths of a millimeter up to a few millimeters. They are generally made of a ceramic block covered by a resistive layer between two metallized regions for connection (Figure 13.6). The geometry of the resistive layer is adjusted with a laser during the fabrication process in order to meet tight tolerances. Sometimes two connecting ribbons are attached to facilitate insertion. Resistive loads and attenuators are manufactured in the same manner.

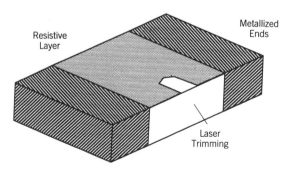

FIGURE 13.6 Discrete resistor.

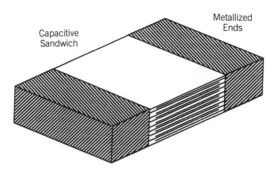

FIGURE 13.7   Discrete capacitor.

13.5.3   Capacitors

Seen from the outside, capacitors look very much like resistors. They consist of a sandwich of ceramic and metal layers (Figure 13.7). Available capacitance values range from fractions of a picofarad up to several nanofarads.

Adjustable capacitors are available for insertion into microstrip circuits, either series or shunt connected (in the latter case, a hole must be drilled through the substrate). They are a few millimeters in size, which limits their use up to a few gigahertz because resonances appear at higher frequencies. These adjustable components are most useful for fine tuning or to compensate for component nonuniformity.

13.5.4   Inductors

Miniature inductors for microstrip consist of helically deposited metal strips on very thin multilayer ceramic substrates connecting one layer to the next at each turn. A multiturn "coil" of very small dimensions is obtained in this manner. Large inductance values with small loss can be realized in a small volume.

13.5.5   Semiconductor Diodes

Most kinds of microwave diodes are available for mounting on microstrip: Schottky, pin, varactors, Gunn, IMPATT, TRAPATT, and others (Chapter 9). Diodes can be inserted in chip form or encased in various packages: leadless inverted device (LID), beam lead (connection with thin strips) or cylindrical cases, with or without mounting screw. The last package is mounted through the substrate, with the lower part in close contact with the ground plane to ensure heat removal.

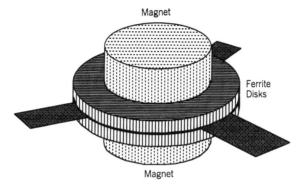

FIGURE 13.8 Circulator.

13.5.6 Transistors

The increasing importance of microstrip circuits is closely related to the availability of transistors, bipolars, MESFETs, and HEMTs that are available for operation at increasingly high frequencies (Chapter 8). Special cases with low parasitics were designed to avoid degrading the performances of the chip. They permit an easy insertion with minimal discontinuities. A hole must be drilled through the substrate to insert power devices, which require a heat sink. The best performances are achieved with chip-mounted transistors.

13.5.7 Circulators and Isolators

Junction isolators and circulators can be mounted on microstrip. They are small ferrite disks, a few tens of millimeters in diameter, with small ribbons for connection within the circuit (Figure 13.8). Holes must be drilled into the substrate to permit insertion. These components are relatively heavy and large due to the presence of permanent biasing magnets (Section 6.3.3).

13.6 BONDING TECHNIQUES

13.6.1 Requirements

The mounting of lumped elements on microstrip substrates requires particular care:

The electrical connections must be made very carefully.

Discontinuities between component and circuit must be kept as small as possible.

The mounting must be mechanically stable.

Precautions are required to avoid excessive heating of delicate components during the mounting procedure.

Connections may deteriorate due to aging (oxidation, etc.).

Since components are generally quite small and light, their mechanical stability is most often provided by the electrical connections themselves.

13.6.2   Soldering

The classical soldering, currently used in electronics, can be used for mounting components on a microstrip substrate, but specific equipment is required to prevent damage of the components. The soldering iron must have a very narrow tip, and a number of solders based on indium, tin, and lead can be used, with melting points from 143 to 280° C [Laverghetta 1991].

13.6.3   Thermocompression Bonding

Thermocompression bonding, which involves applying heat and pressure at the same time, produces an interatomic diffusion and thus a high-quality weld. This process usually utilizes gold as the welding material. The duration of the bonding operation (typically 1–3 s) is a critical parameter of the bonding process.

13.6.4   Ultrasonic Welding

Ultrasonic welding operates on the same physical principle as bonding, namely by heat and pressure. The heat is produced by mechanical rubbing by a point vibrating at ultrasonic frequencies. This technique is particularly well suited for very sensitive components.

13.6.5   Joule Welding

Welding can also be realized by Joule heating, circulating a current between two electrodes. The interest of this process is that heat can be applied very locally. On the other hand, semiconductor devices might be damaged.

13.6.6   Gluing

Devices can also be glued with conductive epoxy loaded with fine metal particles, generally of gold or silver. This process is extremely useful for mounting very delicate components and for encasing microstrip circuits [Laverghetta 1991].

13.6.7 Drilling Holes

Some components require an electrical connection to the ground plane or a good thermal heat sink: to insert them into the circuit, one drills a hole through the substrate. The process is straightforward with synthetic and composite substrates as long as the drill bit and speed are selected correctly. It is much more difficult to drill holes in ceramic substrates, which are hard and brittle. Two techniques can be used:

1. Drilling with an abrasive powder (carborundum) and a rotating tube or an ultrasonically driven point. This approach is time consuming.
2. A laser can realize perfect holes, even non-round ones, in a fraction of a second. However, a special drilling setup is required.

In all situations, one tries to drill as few holes as possible in ceramic substrates. In multilayer structures the internal surface of some holes is metal plated to provide a connection between the two faces of a dielectric layer. Such plated holes are called via holes.

13.7 POSSIBLE ADJUSTMENTS

An important drawback of the printed circuit procedure is that adjustments are difficult or even impossible. The only adjustable devices available for printed circuits are miniature tunable capacitors that can be used at the lower microwave bands (Section 13.5.3). Part of the top conductor may be removed by mechanical scratching or with a laser: this is an irreversible process. Conducting paint may be deposited to complete a missing conductor or to widen a strip (this may actually add some losses). Such operations can be made during the development process but may not be acceptable in a finished product. It is therefore recommended to simulate the operation of the circuit as accurately as possible before building it and to carefully analyze its sensitivity to parameter changes. As a rule, broad-band designs tend to be less sensitive to errors than narrow-bandwidth ones (especially those containing filters).

CHAPTER FOURTEEN

Measurement Techniques

14.1 CIRCUIT MEASUREMENTS

14.1.1 Microwave Measurements

To measure circuits and antennas at microwave frequencies, one compares a signal at some point in space with another signal at the same point (reflection) or some other point (transmission). One determines in this manner power ratios and phase shifts that are related to the terms of the scattering matrix (*S*-parameters, Section 4.3.3). Impedance, admittance, and chain matrices (Section 4.2) can then be determined, and equivalent circuits established (Section 4.5).

When operating at low signal levels, most passive circuits are linear; that is, their parameters do not vary with the power level (Section 4.4.2). It is not necessary to determine actual power levels, only power ratios, and the measurements are called relative. Semiconductor diode detectors operating in the quadratic portion of their *I-V* characteristic are generally used for this purpose (Section 9.2.2) [Watson 1969].

On the other hand, some transistor amplifiers are highly nonlinear even at low signal levels. In this case, the power level of both signals must be completely determined. Harmonic signals are also generated, so power measurements must be carried out at several harmonic frequencies to properly characterize the device. Other microwave devices also become nonlinear at high power levels. For instance, the attenuation of ferrite devices increases with power, while the isolation loss of isolators decreases (Section 6.3). Nonlinear behavior can also result from heating by high-power signals.

Since microstrip circuits and antennas are mostly designed for low-power operation, this chapter is devoted to the measurement of linear devices.

14.1.2 Reflectometry

The reflection at a device's input is one of the most significant parameters characterizing a component. When all other ports are terminated by reflec-

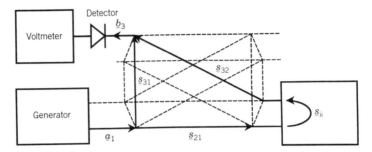

FIGURE 14.1   Simple reflectometer.

tionless loads, the reflection factor at port i of the device is the parameter $\underline{s}_{ii}$ (Section 4.3.8). To measure it, one uses a directional coupler (Section 6.6) as shown in Figure 14.1 [Sucher and Fox 1963].

The terms $\underline{s}_{12}$, $\underline{s}_{13}$, and $\underline{s}_{23}$ refer to the coupler, while the term $\underline{s}_{ii}$ is the reflection factor of the device under test. The signal coming out of port 3 of the coupler is related to the input signal at port 1 by

$$\underline{b}_3 = \underline{a}_1(\underline{s}_{13} + \underline{s}_{12}\underline{s}_{ii}\underline{s}_{23}). \tag{14.1}$$

The device under test is then replaced by a short circuit, and the measurement is repeated, yielding

$$\underline{b}_{3cc} = \underline{a}_1(\underline{s}_{13} - \underline{s}_{12}\underline{s}_{23}). \tag{14.2}$$

Taking the ratio of the two quantities and assuming that the input signal remained constant during the complete procedure (a stable signal source is required), one can extract the value of $\underline{s}_{ii}$:

$$\underline{s}_{ii} = -\frac{\underline{b}_3}{\underline{b}_{3cc}} - \frac{\underline{s}_{13}}{\underline{s}_{12}\underline{s}_{23}}\left(1 - \frac{\underline{b}_3}{\underline{b}_{3cc}}\right). \tag{14.3}$$

The microwave signal at the coupler's output is fed to a diode rectifier (Section 9.2) which detects the amplitude but not the phase of the microwave signal. The amplitude of the reflection factor is thus only known with some uncertainty, located between two bounds (neglecting $\underline{b}_3/\underline{b}_{3cc}$ in the error term):

$$\left|\left|\frac{\underline{b}_3}{\underline{b}_{3cc}}\right| - \left|\frac{\underline{s}_{13}}{\underline{s}_{12}\underline{s}_{23}}\right|\right| \le |\underline{s}_{ii}| \le \left|\frac{\underline{b}_3}{\underline{b}_{3cc}}\right| + \left|\frac{\underline{s}_{13}}{\underline{s}_{12}\underline{s}_{23}}\right|. \tag{14.4}$$

The error term is due to the leakage resulting from the coupler's finite directivity (Eq. 6.51). The range of uncertainty narrows as the directivity increases. A 40-dB-directivity coupler yields an error range of $\cong \pm 0.01$, while a directivity of 60 dB is required to reduce the error range to $\cong \pm 0.001$. A coupler with a directivity of only 20 dB is practically useless for reflection measurements. It is not possible to correct the error or reduce the uncertainty because the rectifier diode in the test setup of Figure 14.1 does not provide any information on the phases of the microwave signals.

To determine the phase of the reflection factor, one must add a second coupler which samples a part of the incident signal. The outputs of the two couplers are then combined in a hybrid junction or coupler to make an interferometric bridge [Sucher and Fox 1963].

14.1.3   Network Analyzers

To completely measure the reflection and transmission parameters of a two-port device, one connects two couplers at each port, sampling the ingoing and outgoing signals (Figure 14.2). Four measurement channels are obtained in this manner and are connected to the inputs of a network analyzer.

The network analyzer shifts the four signals down in frequency, generally in two steps. Sampling heads bring the signals down to a fixed intermediate frequency in the 20–30 MHz range, and then diode mixers (Section 9.3) provide low-frequency signals in the 200–300 kHz range. The relative amplitudes and phases of the signals are maintained during the two operations. The low-frequency signals obtained are compared by electronic means, which determine their amplitude ratios and phase shifts. When the generator is connected to port 1 of the device under test, the ratios $\underline{b}_2/\underline{b}_1$ and $\underline{b}_3/\underline{b}_1$ are determined, whereas the ratios $\underline{b}_3/\underline{b}_4$ and $\underline{b}_2/\underline{b}_4$ are measured when it is connected to port 2. The four quantities are related, respectively, to the

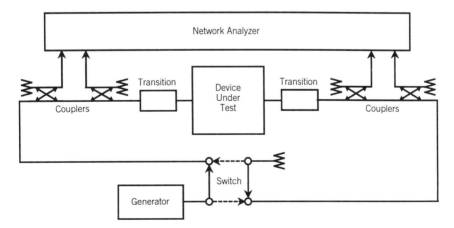

FIGURE 14.2   Network analyzer.

parameters $\underline{s}_{11}$, $\underline{s}_{21}$, $\underline{s}_{22}$, and $\underline{s}_{12}$ of the device, to which are, however, added error signals contributed by the nonideal character of the surrounding circuitry.

Traditionally, errors used to be physically suppressed with matching adapters and tuners [Engen and Beatty 1959]. When errors can be completely canceled, the measured ratios yield the *S*-parameters. Unfortunately, matching is time consuming and tedious and requires considerable skill and care. Because it must also be repeated for every signal frequency, this approach is clearly impractical for everyday measurements. It is nearly only used in national calibration and certification laboratories, where dedicated measurements setups remain permanently tuned at specified frequencies.

14.1.4 Calibration Process

Computer techniques can now determine and correct recurrent errors. The couplers of the setup in Figure 14.2 are not perfectly directive, so spurious signals leak across channels. Discontinuities (connections, etc.) also cause reflections that produce similar results. These effects are systematic, and when the environment remains stable the errors can be evaluated. All effects are included in a 12-term error model, schematically represented in Figure 14.3 by two flow graphs.

The first subscripts of the error terms correspond to directivity (D), source match (S), isolation (X), transmission tracking (T), reflection tracking (R), and load match (L). The two flow graphs correspond to the two positions of the generator switch, defined by the second subscripts as forward (F) and reverse (R) connections. Two different flow graphs are required to account

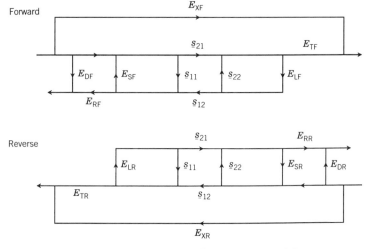

FIGURE 14.3 Two-port 12-term error model.

for possible system imbalance: the switch "sees" different terminating impedances as its position is changed from forward to reverse.

To calibrate the measurement setup, one connects accurately known devices (standards) to the network analyzer. The systematic errors are defined by comparing the measured and known responses of the standards. The 12 terms of the error model (Figure 14.3) are determined from the differences observed. The network analyzer carries out the complete calibration process for all frequencies to be used for the measurements.

For coaxial two-port calibrations, three calibrated impedance references (short circuit, open circuit, and matched load) and a transmission standard are required. The accuracy with which the systematic errors can be characterized depends critically on how well the standards are defined, since imperfections in the calibration standards produce residual effects. Under ideal conditions and with perfectly known standards, systematic errors can be completely characterized and removed.

14.1.5   Through-Reflect-Line Calibration Technique

Since network analyzers are equipped with coaxial ports, microstrip circuits must always be measured across microstrip-to-coaxial transitions (Section 1.5.4). The transitions required to connect the device produce mismatches, phase shifts, and losses that must be separated from the device's own characteristics (Section 4.8). The contribution of connectors and transitions can be canceled by calibrating the network analyzer directly with microstrip standards.

However, in noncoaxial systems such as microstrips it is difficult to build and characterize impedance standards. Microstrip shortcircuits are inductive, open microstrip lines radiate energy (Section 5.2.1), and loads are seldom purely resistive. A transmission line standard is easier to realize, leading to the development of a new calibration procedure—the through–reflect–line (TRL) calibration technique, which is best suited for noncoaxial and dispersive transmission media [Franzen and Speciale 1975; Engen and Hoer 1979].

The TRL calibration technique consists of three main steps, as sketched in Figure 14.4 [Hewlett–Packard 1992]:

1. *Through*. The two halves of the test fixture are directly connected to each other, and measurements of transmission and reflection are carried out for both directions. Reference planes are generally defined at this stage, setting $\underline{s}_{11} = \underline{s}_{22} = 0$ and $\underline{s}_{12} = \underline{s}_{21} = 1$. Proper connection of the upper strip is provided by a thin conductor at the end of a dielectric rod. Particular care is required to ensure the repeatability of connections.

2. *Reflect*. The two halves of the test fixture are separated, providing large and identical reflections. The reflection factors are measured on both sides. It is not necessary to know the actual value of the reflection amplitude, but the range of its phase must be specified (within $\pm 90°$). It is also possible to

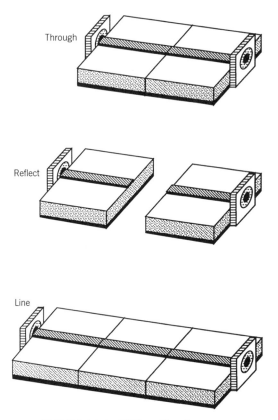

FIGURE 14.4 TRL calibration steps.

terminate the two halves with short circuits as long as both sides are identical.

3. *Line*. The calibrated section of transmission line is inserted between the two halves of the test fixture, and measurements of transmission and reflection are carried out for both directions. The characteristic impedance of the section must be known precisely because it establishes the impedance reference for all the measurements. The electrical length of the transmission line section may vary between 20° and 160°, the optimal value being 90° ($\lambda_g/4$). The calibration process can, therefore, only cover a limited bandwidth, from a lower frequency f_0 to an upper frequency $8f_0$. It is not necessary to know the line attenuation, but the range of its insertion phase must be specified (within $\pm 90°$). During both the through and line steps, the input signals for the two switch positions (forward and reverse) are compared to account for the inbalance of the test system. Finally, the forward and reverse isolations are measured as leakage between the ports with each port terminated.

Altogether, the calibration process involves 16 measurements, from which the 12 terms of the error model (Figure 14.3) are determined. The mathematical development is, however, quite involved [Engen and Hoer 1979; Rubin 1990]. The additional information provided by the measurements is used to determine the phase and amplitude of the reflections (reflection step) and the electrical length and attenuation of the transmission line standard (line step).

It is also possible to use two calibrated sections of transmission line, one for the through step and the other one for the line step. The lengths of the two sections must then be different, and one of the two is selected to specify the reference planes. It is also possible to use matched standards to specify the impedance level.

14.1.6 Measurements and Correction of Errors

The device to be tested is then inserted into the test fixture used for the calibrations, which is connected to the network analyzer, and measurements are carried out. Accurate values of $\underline{s}_{11}$, $\underline{s}_{21}$, $\underline{s}_{22}$, and $\underline{s}_{12}$ are extracted from the measured data by calculating the error corrections with a deembedding process (Section 4.8), based on the correcting terms determined by the calibration. The reflection factors $\underline{s}_{ii}$ are generally displayed on a Smith chart (Section 4.6.6).

The range of errors that can be corrected remains bounded. Even the best calibration process cannot provide accurate results when transitions and connectors are badly mismatched or lossy. An easy way to check the accuracy is to compare the values determined for $\underline{s}_{21}$ with those for $\underline{s}_{12}$, since the two must be identical for all reciprocal devices.

14.1.7 Antenna Measurements

An antenna measurement requires two antennas: one to transmit the signal, the other to receive it. One of the two is the antenna to be tested. The ratio of received power to transmitted power, and sometimes also the phase shift, is determined, first, when the antennas are pointed toward each other and, second, as a function of the direction of pointing. A network analyzer can be used to measure antennas (Figure 14.5).

In the close vicinity of the antenna, the fields exhibit complex behavior, since reactive components are present in addition to the radiated field. As one moves away from the antenna, the reactive components die out and only the radiated field remains in the Fraunhofer, or far-field, region. The distance R between the two antennas must satisfy the condition

$$R \geq 2D^2/\lambda, \tag{14.5}$$

where D is the largest dimension of the antenna and λ is the wavelength.

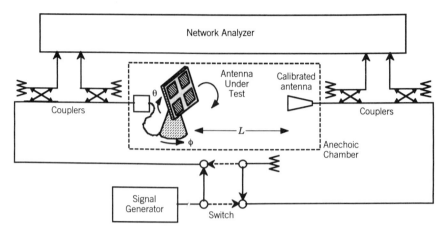

FIGURE 14.5   Antenna test setup.

If two antennas were to radiate uniformly over the entire space, power would be evenly distributed in all directions and the power ratio would be $(\lambda/4\pi R)^2$. But actual antennas always concentrate the signal toward some preferential directions and are thus called directive. This characteristic is represented by the power gain $G(\theta, \phi)$, which is a function of the angles defined with respect to the antenna. The power ratio is then

$$P_\mathrm{r} = P_\mathrm{t} G_1 G_2 (\lambda/4\pi R)^2, \tag{14.6}$$

where P_r is the received power, P_t is the transmitted power, and G_1, G_2 are the antenna gains. This expression is reciprocal, either antenna can be used as the transmitter and the other as the receiver.

From the measurement of the power ratio, one can determine the product $G_1 G_2$ but not the two quantities separately. More generally, the parameters of one antenna are known. It is also possible to have two identical antennas, in which case $G_1 = G_2$. Finally, one may use three antennas and determine the gains of the three from the three gain products measured.

Antenna gain takes into account its directivity D and the power losses in the antenna itself, represented by the antenna efficiency η:

$$G = \eta D. \tag{14.7}$$

Theoretical derivations generally provide the directivity (gain of an ideal lossless antenna) while measurements yield the gain.

14.2   SUBSTRATE PROPERTIES

14.2.1   Introduction

Microstrip circuit design requires accurate knowledge of the substrate's characteristics; especially for frequency-sensitive components such as filters (Sections 7.3 and 7.4). The specifications given by the manufacturer for permittivity and thickness are often too broad for design purposes (Table 13.1). It is therefore necessary to actually measure this parameter for the substrate used, and several techniques were developed for this purpose [Traut 1989a].

14.2.2   Stripline Resonator Method

Unclad substrates are clamped between metallic plates, with ground plane foils and a permanent pattern card bearing probe lines and a rectangular resonator [IPC 1988]. A clamping force is applied to ensure proper contacts. The resonant frequency and quality factor are measured with standard resonant cavity measurement techniques [Sucher and Fox 1963]. The real part of the relative permittivity ε' and the dissipation factor $\tan \delta$ are simply

$$\varepsilon' = \left(\frac{nc}{2f_{\mathrm{r}}(L + \Delta L)} \right)^2, \quad \tan \delta = \frac{1}{Q} - \frac{1}{Q_{\mathrm{c}}}, \tag{14.8}$$

where c is the velocity of light, L is the length of the resonator, ΔL is the length correction to account for end effects in stripline, n is the number of half-wavelengths along the resonator, Q is the measured quality factor, and Q_{c} is the quality factor contributed only by conductor losses (to be determined from the known conductivity and dimensions of the conductors). By using several test patterns with different resonator lengths, one determines the effects of the ends and of the coupling. This method can only be used with unclad substrates.

14.2.3   Microstrip Resonator Method

In the microstrip resonator approach, a resonator and coupling lines are realized by the photolithographic process on the substrate to be measured (Section 13.3). The resonator is then placed in a test fixture, where the resonant frequency and the quality factor are determined. A good ohmic connection must be ensured between the ground plane of the microstrip and the test fixture. Since at resonance a patch tends to radiate, it is recommended to place a cover above the resonator, but not too close, in order to not perturb the measurements (Section 3.4.10).

A rectangular patch resonator can be used to measure the substrate properties (Section 7.1.2). The resonant frequency is given by Eq. 7.1. The effective permittivity ε_e can be extracted from this equation, but the length correction factor ΔL is a complex function of the unknown ε_e and the unknown ε_r (Section 5.2.1). The determination is therefore complicated, and approximate formulations were established for this purpose [Traut 1989a]. It is also possible to insert an analysis program within an optimization loop (Figure 12.1).

End effects and some radiation loss can be avoided with a circular resonant ring (Section 7.1.4). The computer program Epsilon was developed to design and cut the mask, with the technique developed for the circuit layout software Micros (Section 12.4.5). When the resonator has been realized and measured, the values measured are introduced in the program, which then calculates the permittivity [Zürcher et al. 1986].

The microstrip resonator method determines the effective permittivity under actual operating conditions, since the procedure used to realize the resonator and the coupling lines is exactly the same one that will be used to realize the circuit itself. The technique is destructive, since a part of the material must be used to make the resonator.

14.2.4   Full-Sheet Test Method

A rectangular sheet of microstrip material clad with metal on both sides but not on the edges is treated as a parallel-plate resonator in which the electric field is assumed to be simply normal to the metal layers. Probe connections are made at edges or corners of the plate, and the frequency response is measured. A first approximation for the resonant frequencies, assuming that the edges are perfect magnetic conductors, is

$$f_r \cong \frac{c_0}{2\sqrt{\varepsilon_r}} \sqrt{\left(\frac{m}{a}\right)^2 + \left(\frac{n}{b}\right)^2}, \qquad (14.9)$$

where m and n are the resonant mode indices, indicating the number of half-wavelengths in the two transverse directions, while a and b are the resonator dimensions. The responses of modes in the measured response of the resonator are then identified with (m, n) pairs, and the value of ε_r is determined from Eq. 14.9. The latter operation presents some difficulties when large structures are measured, because the spectrum of resonant modes may become too dense around the frequency of operation so that it may no longer be possible to unambiguously identify a specific mode (just as in a microwave oven).

In a similar approach, the dielectric is completely metallized, even around the edges, leaving small apertures for coupling [Ladbrooke et al. 1973]. In this case the edge boundaries become electric walls, and the foregoing procedure is also used for the measurements.

14.2.5 Cavity Perturbation Method

In the standard approach to measure dielectric materials, a small sample of the dielectric material is cut and placed in a resonant cavity, generally at the center [Sucher and Fox 1963]. The electric field of the resonant mode selected must reach its maximum value at the sample location. The permittivity is then approximately given by the perturbation expression.

$$f_r - f_{r0} + 2j\frac{f_{r0}}{Q_0} \cong -\frac{f_r}{2} \frac{\int_{\Delta V}(\varepsilon_r - 1)|\mathbf{E}_0|^2\, dV}{\int_V |\mathbf{E}_0|^2\, dV}, \qquad (14.10)$$

where $\mathbf{E}_0$ is the electric field in the unloaded cavity, V and ΔV are, respectively, the volumes of the cavity and of the sample, f_r and f_{r0} are the resonant frequencies of the cavity with and without the sample, and Q_0 is the quality factor. The integrals must be evaluated for the shape, size, and resonant mode of the cavity [Gardiol 1984].

14.2.6 Remarks

In previous sections it was implicitly assumed that the substrate thickness remains constant and that the permittivity does not vary across the sheet of microstrip material tested. When these conditions are satisfied, the results of the measurements can be used to characterize the complete sheet.

14.3 NEAR FIELDS

14.3.1 Introduction

The measurements of microwave circuits (Section 14.1) only provide information related to signals at some discrete points in space, such as their reflection at a port of a device, and their transmission between two ports. This information does not "show" what actually takes place within the device, how signals travel along lines and cross junctions, and how they "mysteriously" disappear at isolated ports.

The radiation pattern of antennas is measured in the far field (Section 14.1.7) over some particular planes. It does not indicate which edges of a printed antenna predominantly radiate, which are inactive, and which may contribute spurious radiation (cross-polarization, side lobes).

Near-field techniques were developed to measure antennas, particularly under situations where the condition of Eq. 14.5 cannot be satisfied. Near-field measuring apparatus is now available, but it tends to be highly specialized and expensive. The modulated scatterer approach, proposed by Richmond [1955] and developed by Hygate and Nye [1990, 1991], makes it possible even for modestly equipped laboratories to measure the near field and, thus, to obtain valuable information on the actual operation of printed circuits and antennas.

14.3.2 Measurement Principle

In principle, near-field measurements could be made with a network analyzer and a probe (Section 14.1.3). However, the probe must be connected, and a coaxial cable close to the dielectric interface would perturb the field even if the cable were very thin. In addition, the phase stability of a thin cable is generally insufficient to make accurate measurements.

In the modulated scatterer technique, the probe used to explore the field does not carry a microwave signal, but instead contains a diode modulated by a low-frequency signal. This probe locally perturbs the field and produces a modulated reflection. At the receiver, the modulated reflection is extracted from the background reflected signal, which is unmodulated. A homodyne receiver and a lock-in amplifier provide a very sensitive system [Hygate and Nye 1990].

The measurement setup is shown in Figure 14.6. A stable microwave source (for instance, a sweep oscillator phase-locked to a microwave counter) provides a signal, which is then amplified to a level adequate to drive the mixers of the homodyne receiver. An automatic level control (ALC) loop stabilizes the power level. The signal is fed to the device under test (DUT), mounted on a standard plotter. A movable arm made of low-permittivity foam carries the probe, modulated by a low-frequency (10-kHz) generator, which also feeds the lock-in amplifier. The two outputs of the homodyne receiver are sequentially switched to the input of the lock-in amplifier, with a digital meter measuring the output signal.

In the homodyne receiver a sample of the input signal is mixed, in phase and in quadrature, with the rf signal returning from the DUT. The two if

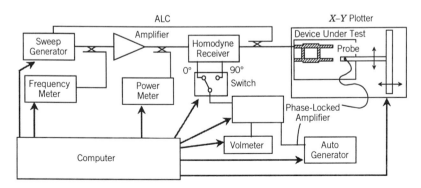

FIGURE 14.6 Schematic diagram of a near-field measurement setup.

outputs carry the 10-kHz signals with a 90° phase offset, from which the amplitude and the phase of the field component can be computed. The homodyne receiver is built on a single microstrip circuit, with identical high-quality double-balanced mixers, to ensure a high symmetry of the structure and, thus, suitable sensitivity and stability.

The probe is moved by the plotter over the area specified by the user, with the required resolution, and both output signals (0° and 90°) from the homodyne receiver are recorded for each point. The measured field can then be displayed or analyzed by the computer. The system developed measures from two to three points per second, with a dynamic range larger than 40 dB [Zürcher 1992].

14.3.3 The Probes

Several probes were realized and tested, where use was made of low-cost low-frequency miniature diodes. The diodes are modulated directly rather than by optical means because this provides a higher dynamic range. The use of high-resistivity wires, together with absorbers and ferrite beads, prevents the propagation of the microwave signal on the low-frequency circuit. It was found experimentally that the wires have practically no influence on the measurements.

Probes for electric field (dipole type) and for magnetic field (loops) were realized with dimensions between $\lambda/20$ and $\lambda/50$. All six field components $(E_x, E_y, E_z, H_x, H_y,$ and $H_z)$ can be measured.

14.3.4 Example for a Microstrip Circuit Element

The amplitude of the normal E_z field component on a branch line coupler (Section 6.5.1) is represented in Figure 14.7. The field maximum is located close to the input port and evenly divided between the two outputs. The most interesting feature is the cancellation of the fields in the isolated port. Other

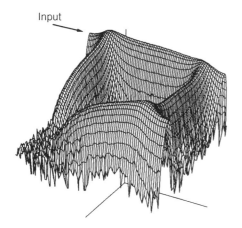

Input

FIGURE 14.7 Amplitude of the E_z field component on a 3-dB branch line coupler.

representations, using a gray scale, are given in Figure 6.15 (branch line coupler) and 6.18 (rat race hybrid ring).

14.3.5   Example for a Microstrip Antenna

Figure 14.8 represents, at three frequencies, the amplitude of the H_z field component on a patch antenna with parasitic elements on both sides (Section 11.4.2). The field is mostly concentrated near the outer parasitic elements at the low frequency (1.7 GHz). The main contribution at the center frequency (1.8 GHz) is provided by the inner parasitic elements. At the high frequency (1.9 GHz), the center patch resonates and the field is concentrated on its edges.

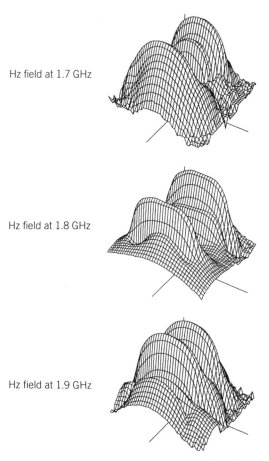

Hz field at 1.7 GHz

Hz field at 1.8 GHz

Hz field at 1.9 GHz

FIGURE 14.8 Amplitude of the H_z field component at three frequencies on a patch antenna with parasitic elements.

Solutions To The Problems

CHAPTER 1

1.6.1 The period of a 465-MHz signal is simply obtained by taking the inverse:

$$T = 1/465 \times 10^6 = 2.1505 \times 10^{-9} \text{ s} = 2.1505 \text{ ns.}$$

Its wavelength in free space is

$$\lambda_0 = c_0/f = c_0 T = 2.998 \times 10^8/465 \times 10^6 = 0.6447 \text{ m.}$$

1.6.2 The frequency of a signal with a free-space wavelength of 22 cm is
a. In microwaves

$$f = 2.998 \times 10^8/0.22 = 1.36272 \times 10^9 \text{ Hz} = 1.36272 \text{ GHz.}$$

b. In acoustics

$$f = 300/0.22 = 1363 \text{ Hz} = 1.363 \text{ kHz.}$$

1.6.3 The reduction in size is approximately proportional to the square root of the permittivity ratio; that is,

$$\sqrt{2.3/80} = 0.1695.$$

Actually, the reduction is proportional to the effective permittivity ratio, which also depends on the geometry (see Problem 3.8.10).

CHAPTER 2

2.7.1 An electric field of 100 V/m acting on a charge of 5 coulombs produces a force of $100 \times 5 = 500$ VA-s/m $= 500$ N.

2.7.2 When the electric field of a wave is given by the expression $\mathbf{E} = \mathbf{e}_x \sin(\omega t - \beta y)$, with ω and β constants, the magnetic field in air is obtained by applying the first Maxwell curl equation (2.4), which after simple calculations yields

$$\mathbf{H} = -\mathbf{e}_z \frac{\beta}{\omega \mu_0} \sin(\omega t - \beta y).$$

The electric charge density is obtained by taking the divergence, which is here equal to zero.

2.7.3 For the electric field $\mathbf{E} = \mathbf{e}_x \sin(\omega t - \beta x)$, with constants ω and β, the curl vanishes, so there is no magnetic field. The electric charge density is given by the divergence, yielding $\rho = \beta \varepsilon_0 \cos(\omega t - \beta y)$. This situation corresponds to an electrodynamic wave in a plasma.

2.7.4 The curl of $\mathbf{B} = \mathbf{e}_z \sin \omega t \cos(\pi x/a) \cos(\pi z/d)$ is not equal to zero, this expression cannot represent an induction field.

2.7.5 The conduction current in copper is equal to the displacement current when $\sigma = \omega \varepsilon_0$. Solving for the frequency, we find $f = 1.045 \times 10^{18}$, well above the frequency ranges at which microstrips can be used.

2.7.6 The phasor–vector $\underline{\mathbf{E}} = \mathbf{e}_z + j\mathbf{e}_x$ corresponds in the time domain to the electric field $\mathbf{E} = \sqrt{2}(\mathbf{e}_z \cos \omega t - \mathbf{e}_x \sin \omega t)$. The amplitude of the field is a constant $\sqrt{2}$, since the end of the electric field vector describes a circle in the x, z plane.

2.7.7 Considering the expression of the curl in rectangular coordinates and comparing the terms, it appears that $\mathbf{B} = yz\,\mathbf{e}_z - xy\,\mathbf{e}_x$ can derive from the vector potential $\mathbf{A} = xyz\,\mathbf{e}_y$. But this is not the only possible solution, as any curl-less function can be added to it. Since the curl of a gradient is identically zero, we can write the general solution in the form

$$\mathbf{A} = xyz\,\mathbf{e}_y + \nabla \Phi,$$

where Φ is any arbitrary function of x, y, and z.

2.7.8 For plane waves traveling in air, the expression $\gamma \cdot \gamma$ in Eq. 2.41 must be purely real. Developing $\gamma = \alpha + j\beta$, we find that we must have $\alpha \cdot \beta = 0$. This condition can be satisfied by letting $\alpha = 0$, in which case we have uniform plane waves, or when $\alpha \perp \beta$, in which case the plane waves are nonuniform. This behavior is encountered in surface waves (Section 1.3.5) and in leaky waves (Section 1.3.4).

2.7.9 The skin depth and the metal impedance of copper are obtained from Eq. 2.42 and 2.43, with $\sigma = 58.13 \times 10^6$ S/m, yielding at 30 GHz:

$$\delta = 3.811 \times 10^{-7}, \qquad \underline{Z}_m = 0.0514(1 + j).$$

CHAPTER 3

3.8.1 Using Wheeler's relationships (Section 3.4.8), we find for a line with a characteristic impedance of 80 Ω on a polypropylene substrate ($\varepsilon'_r = 2.18$): $w/h = 1.419$ and, since $h = 0.8$ mm, $w = 1.135$ mm.

3.8.2 Analysis equations provide the effective permittivity ε_e and the characteristic impedance Z_c. Two sets of equations were used, yielding, respectively,

$$\varepsilon_e = 1.782, \qquad Z_c = 80.29 \ \Omega,$$
$$\varepsilon_e = 1.788, \qquad Z_c = 80.06 \ \Omega.$$

While the calculated value for the impedance is not exactly 80 Ω, the difference observed is less than 1%.

3.8.3 The extremal values are obtained for the following combinations:
a. Smallest value of $\varepsilon_r (= 2.13)$ and largest value of $h (= 0.84)$, in which case

$$Z_c = 83.12 \ \Omega, \qquad \varepsilon_e = 1.745.$$

b. When ε_r takes its largest value ($= 2.23$) and h its smallest one:

$$Z_c = 77.43 \ \Omega, \qquad \varepsilon_e = 1.820.$$

3.8.4 An error of $\pm 10\%$ on the width of the upper conductor produces the following changes:

a. -10% $w = 1.0218$ mm, $Z_c = 84.73 \ \Omega$, $\varepsilon_e = 1.7729$,
b. $+10\%$ $w = 1.2489$ mm, $Z_c = 76.32 \ \Omega$, $\varepsilon_e = 1.7902$.

The major effect appears in the characteristic impedance; the change is about 5%.

3.8.5 A strip of width $w = 1.135$ mm with a thickness $b = 50$ μm is equivalent to a strip of zero thickness that has an effective width $w_e = 1.2412$ mm. To compensate for this, calculations are carried out to obtain an effective width of $w_e = 1.135$ mm, found to correspond, after some iterations, to a strip width of $w = 1.031$ mm.

3.8.6 The attenuation produced by substrate losses $(\tan \delta = 0.0003)$ is $\alpha_d = 0.2958$ dB/m at 10 GHz.

3.8.7 The conductor losses for copper $(\sigma = 58.13 \times 10^6$ S/m) produce the following attenuations, in dB/m:

f(GHz)	1	3	5	10
α_m	0.7005	1.2133	1.5664	2.2152

3.8.8 Taking into account dispersion, the effective permittivity at 12 GHz becomes $\varepsilon_{ef} = 1.8171$, instead of $\varepsilon_{es} = 1.7884$ for the quasi-static approximation. To compensate for it, the strip width is adjusted. Calculations yield $w/h = 1.4488$, and therefore $w = 1.1590$ (instead of $w = 1.1353$).

3.8.9 The frequency at which 1% of the power is radiated by an open microstrip line is approximately $f_{1\%} = 3.25$ GHz.

3.8.10 On a substrate with $\varepsilon_r = 80$, an impedance of 80 Ω is achieved with $w/h = 0.001318$, that is, a width of $w = 0.001055$ mm. The effective permittivity is then $\varepsilon_e = 42.615$. The reduction in line width is thus 1076, while the lengths are reduced by $(42.615/1.782)^{1/2} = 4.89$. In the present case, the strip width is extremely narrow and may be difficult to fabricate. If high-permittivity dielectrics are used, it is best to operate with lower characteristic impedances.

CHAPTER 4

4.9.1 The matrix representations of the two-port device are

$$[Z] = \begin{bmatrix} j35 & -j75 \\ -j75 & j35 \end{bmatrix}, \qquad [Y] = \begin{bmatrix} j0.00795 & j0.01705 \\ j0.01705 & j0.00795 \end{bmatrix},$$

$$\begin{bmatrix} A & B \\ C & D \end{bmatrix} = \begin{bmatrix} -0.4667 & -j58.65 \ \Omega \\ -j0.01333 \ \text{S} & -0.4667 \end{bmatrix}.$$

The series impedance of the T network is $-j40$ Ω, the shunt impedance $-j75$ Ω.

The series admittance of the Π network is $-j0.01705$ S, while the shunt admittance is $j0.025$ S.

4.9.2   The impedance matrix is given by eq. 4.16, taking into account the fact that the lines on both sides have the same characteristic impedance:

$$[\underline{Z}] = \left\{ \begin{bmatrix} 1 & 0 \\ 0 & 1 \end{bmatrix} + \begin{bmatrix} 0.6 & -j0.8 \\ -j0.8 & 0.6 \end{bmatrix} \right\}$$

$$\times \left\{ \begin{bmatrix} 1 & 0 \\ 0 & 1 \end{bmatrix} - \begin{bmatrix} 0.6 & -j0.8 \\ -j0.8 & 0.6 \end{bmatrix} \right\}^{-1} 50$$

$$= \begin{bmatrix} 1.6 & -j0.8 \\ -j0.8 & 1.6 \end{bmatrix} \begin{bmatrix} 0.5 & -j \\ -j & 0.5 \end{bmatrix} 50 = \begin{bmatrix} 0 & -j100 \\ -j100 & 0 \end{bmatrix}.$$

4.9.3   The device corresponding to the first scattering matrix is reciprocal, lossless, but matched at one port only. It is a symmetrical power divider (Section 6.2.1).

The second device is nonreciprocal, not lossless (but almost lossless), and not matched (but almost matched). It is a nonideal three-port circulator (Section 6.3).

The third device is reciprocal, matched, and lossy: it is a resistive three-port power divider (Section 6.2.4).

4.9.4   The equivalent circuits and matrices for 50-Ω lossless transmission line segments are given below in table form.

Some precautions are required in the determination of the network series and shunt impedances, since there are indeterminations of the form $\infty - \infty$ that must be resolved by taking the limit $\beta L \to \pi$.

	length βL	$\lambda/8$ $\pi/4$	$\lambda/4$ $\pi/2$	$\lambda/2$ π
Impedance	$\underline{Z}_{11} = \underline{Z}_{22}$	$-j50$ Ω	0	$j\infty$
matrix	$\underline{Z}_{12} = \underline{Z}_{21}$	$-j70.7$ Ω	$-j50$ Ω	$j\infty$
Chain	$\underline{A} = \underline{D}$	0.7071	0	-1
matrix	$\underline{B}$	$j35.5$ Ω	$j50$ Ω	0
	$\underline{C}$	$j0.0014$ S	$j0.02$ S	0
T network	series $\underline{Z}$	$j20.7$ Ω	$j50$ Ω	0
	shunt $\underline{Z}$	$-j70.7$ Ω	$-j50$ Ω	$j\infty$
Π network	series $\underline{Z}$	$j35.36$ Ω	$j50$ Ω	0
	shunt $\underline{Z}$	$-j120.7$ Ω	$-j50$ Ω	$j\infty$

4.9.5 Considering the T equivalent circuit of a lossy transmission line of length L and taking the limit $L \to dz$, the series and shunt impedances tend to those of the infinitesimal length of line as

$$2\underline{Z}_{\text{series}} = 2\underline{Z}_{\text{c}} \tanh \frac{\gamma L}{2} \to 2\underline{Z}_{\text{c}} \frac{\gamma \, dz}{2} = \sqrt{\frac{\underline{Z}'}{\underline{Y}'}} \sqrt{\underline{Z}'\underline{Y}'} \, dz = \underline{Z}' \, dz,$$

$$\underline{Y}_{\text{shunt}} = \underline{Y}_{\text{c}} \sinh \underline{\gamma} L \to \underline{Y}_{\text{c}} \underline{\gamma} \, dz = \sqrt{\frac{\underline{Y}'}{\underline{Z}'}} \sqrt{\underline{Z}'\underline{Y}'} \, dz = \underline{Y}' \, dz.$$

The development for the Π network is quite similar.

4.9.6 The load can be matched by connecting a section of lossless line of 50-Ω characteristic impedance and a reactive element as follows:

	Length of line	Reactance
Series inductance	0.0897λ	− 49.37 Ω
Series capacitance	0.4219λ	− 49.37 Ω
Shunt inductance	0.1719λ	+ 38.29 Ω
Shunt capacitance	0.3397λ	− 38.29 Ω

4.9.7 This load can be matched either with a quarter-wave transformer or with a generalized transformer.

When using a quarter-wave transformer, a section of line of characteristic impedance $Z_{\text{c}} = 71.2$ Ω of length 0.187λ is first connected to the load. At its input, the impedance is real and is matched by a quarter-wave transformer ($\lambda/4$ section of line) of characteristic impedance 151 Ω.

A generalized transformer has a length of 0.063λ and a characteristic impedance of 15.9 Ω.

In this case, matching with a generalized transformer takes far less space than the traditional quarter-wave transformer. The characteristic impedances of the two devices are also quite different. Note, however, that some loads cannot be matched with a generalized transformer.

CHAPTER 5

5.6.1 The apparent increase in length of an open microstrip line is given by Eq. 5.1, which requires the w/h ratio and the effective permittivity, which must first be determined. The values for 50-Ω lines on substrates 0.6 mm thick are as follows:

ε_r	w/h	ε_{eff}	$\Delta l/h$	Δl [mm]
1.5	3.8949	1.375	0.546	0.327
2	3.2687	1.732	0.494	0.296
3.5	2.2590	2.755	0.449	0.269
10	0.9560	6.758	0.308	0.185
16	0.5776	10.223	0.263	0.158

5.6.2 This problem is treated with the information of Section 5.3. The inductance per unit length of a transmission line is obtained from Eq. 3.26, yielding $L' = Z_c/c = Z_c\varepsilon_e/c_0$. For the two lines (thickness 0.5 mm, relative permittivity 3.5) the w/h ratio and the effective permittivity are given by

Z_c	w/h	ε_{eff}	$L'(\mu\text{H/m})$
80 Ω	0.9463	2.5881	0.429
50 Ω	2.259	2.7475	0.276

The value of L is then

$$L = 0.0845 \text{ nH}, \qquad L_1 = 0.0514 \text{ nH}, \qquad L_2 = 0.0331 \text{ nH}.$$

The capacitance is given by Eq. 5.9: $C = 12.68$ fF.

5.6.3 The ratio w/h is $0.9/0.4 = 2.25$; Eq. 5.12 yields $s/d = 0.551$. Figure 5.5 shows that $d = \sqrt{2}\,w$, so $s = 0.7$ mm.

CHAPTER 6

6.8.1 The scattering matrix of a lossless reciprocal three-port matched at two ports is

$$\begin{pmatrix} 0 & \underline{S}_{12} & \underline{S}_{13} \\ \underline{S}_{12} & 0 & \underline{S}_{23} \\ \underline{S}_{13} & \underline{S}_{23} & \underline{S}_{33} \end{pmatrix}.$$

The losslessness conditions are

$$|\underline{s}_{12}|^2 + |\underline{s}_{13}|^2 = 1, \qquad\qquad \underline{s}_{13}\underline{s}_{23}^* = 0,$$
$$|\underline{s}_{12}|^2 + |\underline{s}_{23}|^2 = 1, \qquad\qquad \underline{s}_{12}\underline{s}_{23} + \underline{s}_{13}\underline{s}_{33}^* = 0,$$
$$|\underline{s}_{13}|^2 + |\underline{s}_{23}|^2 + |\underline{s}_{33}|^2 = 1, \qquad \underline{s}_{12}\underline{s}_{13}^* + \underline{s}_{23}\underline{s}_{33}^* = 0.$$

Comparing the two upper-left expressions, one sees that $|\underline{s}_{13}| = |\underline{s}_{23}|$. However, the product of these two terms equals zero (top right), so both actually vanish. The only nonzero terms are then $|\underline{s}_{12}| = |\underline{s}_{33}| = 1$; that is, the three-port is divided into a lossless matched two-port made with ports 1 and 2 and a totally reflecting one-port at port 3. The two devices are completely uncoupled.

6.8.2   For a power divider with a power ratio of 2.5:1, the characteristic admittances of the two output lines are in the ratio $Y_2/Y_3 = 2.5/1$, with $Y_2 + Y_3 = Y_1 = 0.02$ S. We then have $Y_2 = 0.01428$ and $Y_3 = 0.00572$; the corresponding impedances are $Z_2 = 70\ \Omega$ and $Z_3 = 175\ \Omega$ (this last value may be difficult to realize on microstrip). The scattering matrix is

$$\begin{pmatrix} 0 & 0.845 & 0.534 \\ 0.845 & -0.286 & 0.45 \\ 0.534 & 0.45 & -0.714 \end{pmatrix}.$$

The reader may check that the device is lossless.

6.8.3   A matched n-port resistive power divider is obtained by connecting series resistors between each port and the center of the junction, the resistor value being $[(n-2)/n]\underline{Z}_c$. The scattering matrix then takes the form

$$\frac{1}{n-1} \begin{pmatrix} 0 & 1 & 1 & \cdots & 1 \\ 1 & 0 & 1 & \cdots & 1 \\ 1 & 1 & 0 & \cdots & 1 \\ \cdots & \cdots & \cdots & \cdots & \cdots \\ 1 & 1 & 1 & \cdots & 0 \end{pmatrix}.$$

The power transmitted to each output port is $(n-1)^{-2}$.

6.8.4   This connection is best shown by a picture (Figure S.1).

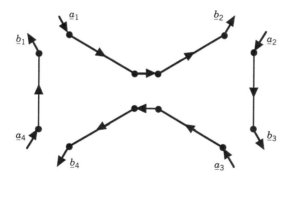

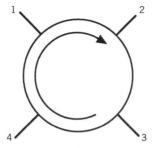

FIGURE S.1   Four-port circulator (connection of two three-port circulators).

6.8.5 The power split of 2.5 : 1 specifies $(\alpha/\beta)^2 = 2.5$, and therefore $\alpha = 0.8452$ and $\beta = -0.5345$. From Eq. 5.34 we find the characteristic impedances of the two lines:

$$Z_1 = Z_c\alpha = 50 \times 0.8452 = 42.26 \ \Omega,$$

$$Z_2 = -Z_c\frac{\alpha}{\beta} = 50 \times \frac{0.8452}{0.5345} = 79.06 \ \Omega.$$

6.8.6 Considering the signal amplitudes and powers in a crossing between two lines of characteristic impedances of 50 Ω and 100 Ω, the scattering matrix is

$$\frac{1}{3}\begin{pmatrix} -1 & \sqrt{2} & 2 & \sqrt{2} \\ \sqrt{2} & -2 & \sqrt{2} & 1 \\ 2 & \sqrt{2} & -1 & \sqrt{2} \\ \sqrt{2} & 1 & \sqrt{2} & -2 \end{pmatrix}.$$

Ports 1 and 3 are at the characteristic impedance of 50 Ω, while ports

2 and 4 are characteristic impedances of 100 Ω. It can readily be checked that the device is lossless.

6.8.7 Equations 6.47 are used to determine the even and odd characteristic admittances of the coupled lines. To have an 18-dB lossless coupler, $|s_2| = 0.9922$ and $|s_3| = 0.1249$. The largest-frequency bandwidth will be obtained in the vicinity of the extremum value, for $\sin \beta L = 1$. The even- and odd-mode admittances are then

$$Y_e = Y_c\sqrt{\frac{1 - 0.1249}{1 + 0.1249}} = Y_c(0.882) \quad \text{and} \quad Y_o = \frac{Y_c}{0.882}.$$

6.8.8 The scattering matrices of three-port and four-port power dividers are obtained directly from Eq. 6.52, removing one column and the corresponding line for a four-port, or two columns and two lines for the three-port. Several possibilities exist, depending on which ports are terminated. One possible choice is given here for both devices:

$$\frac{1}{2}\begin{pmatrix} 0 & 1 & e^{j2\pi/3} \\ 1 & 0 & 1 \\ e^{j2\pi/3} & 1 & 0 \end{pmatrix}, \qquad \frac{1}{2}\begin{pmatrix} 0 & 1 & e^{j2\pi/3} & e^{j2\pi/3} \\ 1 & 0 & 1 & e^{j2\pi/3} \\ e^{j2\pi/3} & 1 & 0 & 1 \\ e^{j2\pi/3} & e^{j2\pi/3} & 1 & 0 \end{pmatrix}.$$

The signals at all output ports have the same amplitude but not the same phase (in resistive dividers the phases are all equal). For the three-ports, the output amplitudes are the same in the terminated five-port and in the resistive divider. In the four-ports, the terminated five-port provides an output amplitude of $\frac{1}{2}$, while that for a four-port divider is only $\frac{1}{3}$.

6.8.9 The scattering matrix of a reciprocal, symmetrical line divider matched at its input and with five outputs is

$$\begin{pmatrix} 0 & B & B & B & B & B \\ B & A & C & C & C & C \\ B & C & A & C & C & C \\ B & C & C & A & C & C \\ B & C & C & C & A & C \\ B & C & C & C & C & A \end{pmatrix}.$$

It is assumed here that A is real, corresponding to a specific selection of reference planes at all outputs. The conditions for which the device is lossless then require that $A = 0.8$, $|B| = 1/\sqrt{5}$, and $C = -0.2$

(real). The output ports are badly mismatched, since a reflection factor of 0.8 corresponds to a VSWR of 9. The phase of $\underline{B}$ is not specified; the reference plane at the input could be selected to have also a real value.

CHAPTER 7

7.6.1 For a line 1.1295 mm wide on a substrate 0.5 mm thick with a relative permittivity $\varepsilon_r = 3.5$, the apparent lengthening was previously determined in Problem 5.6.1, yielding $\Delta l = 0.269$ mm, while $\varepsilon_e = 2.755$. The equivalent patch length is then $15 + 2(0.269) = 15.538$, and the resonant frequencies are given by eq. 7.1 as $m5.816$ GHz.

7.6.2 Introducing the values radius $a = 4$ mm, thickness $h = 0.6$ mm, relative permittivity $\varepsilon_r = 6.5$ into Eq. 7.2, with the effective radius given by Eq. 7.3, yields the resonant frequency of the mode TM_{110} ($\chi_{11} = 1.841$): $f_{110} = 7.743$ GHz.

7.6.3 At 3.27 GHz, the free-space wavelength is 91.74 mm. The median circle of the ring has a circumference of $2\pi R = 40\pi = 125.6637$ mm, which is a full number of loaded wavelengths λ_g. The relative permittivity of the substrate is then

$$\epsilon_{\text{eff}} = \left(\frac{\lambda}{\lambda_g}\right)^2 = n^2 \left(\frac{91.74}{125.6637}\right)^2 = n^2(0.5329).$$

Here we do not know the value of n (there is no "tag" of the resonance indicating what this value is). Quite clearly, we cannot have $n = 1$, since this would correspond to a permittivity lower than 1. Successive values of n then yield the sequence

$$2.132 \quad 4.797 \quad 8.527 \quad 13.324\ldots .$$

Some other indication, for instance information about the type of substrate material, is required to pick up the correct value.

7.6.4 Carrying out the calculations according to Eq. 7.6, the radius of a cylindrical dielectric resonator of relative permittivity $\varepsilon_r = 80$ resonating on its TM_{110} mode at 7.6 GHz is $a = 1.293$ mm.

7.6.5 The scattering parameters of a device made by cascading three transmission lines sections ($\lambda_g/4$ at 3 GHz) with characteristic impedances

10 Ω, 250 Ω, 10 Ω, connected on both sides to 50-Ω lines are

f [GHΩ]	0.2	0.5	0.8	1	1.5	2
$|\underline{s}_{11}|$	0.174	0.449	0.968	0.983	0.999	≈ 1
$|\underline{s}_{12}|$	0.985	0.893	0.251	0.184	0.046	≈ 0

This assembly of line sections behaves like a low-pass filter: attenuation is small at low frequencies and steadily increases with frequency. (Figure S2).

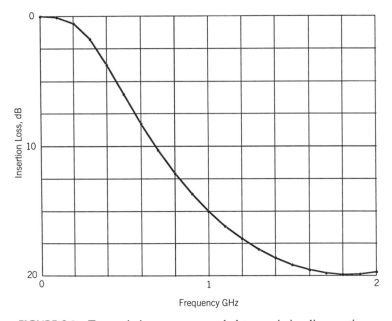

FIGURE S.2 Transmission across cascaded transmission line sections.

7.6.6 If the power divider is matched at its input, there is no solution to this problem. The junction input must be suitably mismatched so that the power reflected by the filter connected at port 2 is exactly compensated for by the intrinsic reflection of the junction.

CHAPTER 8

8.6.1 The power gain of a transistor is obtained by applying Eq. 8.2, which gives

$$\frac{P_{\text{out}}}{P_{\text{in}}} = \left(\frac{f_t}{f}\right)^2 \frac{\text{Re}(\underline{Z}_{\text{out}})}{\text{Re}(\underline{Z}_{\text{in}})} = \left(\frac{12}{10}\right)^2 \frac{51}{10} = 7.344.$$

8.6.2 The cutoff frequency is obtained from Eq. 8.3, in which the 0.25-μm gate length and the saturation velocity of 1.3×10^7 cm/s are introduced:

$$f_c \cong \frac{v_s}{2\pi L_g} = \frac{1.3 \times 10^7 \text{ cm/s}}{2\pi \times 0.25 \,\mu\text{m} \times 10^{-4} \text{ cm}/\mu\text{m}} = 816.8 \text{ GHz}.$$

This is really an extremely high frequency!

8.6.3 A device with an equivalent noise temperature of 40 K and a bandwidth of 200 MHz produces a noise power of

$$N = k_B TB = (1.39 \times 10^{-23})(40)(200 \times 10^6) = 11.12 \times 10^{-14} \text{ W}.$$

This corresponds to -99.5 dBm (dBs with respect to 1 milliwatt).

8.6.4 An input power of 1 W and a power gain of 8 yield an output power of 8 W. With the biasing circuits supplying 18 W, the three efficiencies are respectively

$$\text{total efficiency} \quad \eta_t = \frac{P_{\text{out}}}{P_{\text{in}} + P_b} = \frac{8}{1 + 18} = 0.421,$$

$$\text{partial efficiency} \quad \eta_p = \frac{P_{\text{out}}}{P_b} = \frac{8}{18} = 0.444,$$

$$\text{power added efficiency} \quad \eta_{pa} = \frac{P_{\text{out}} - P_{\text{in}}}{P_b} = \frac{8 - 1}{18} = 0.389.$$

CHAPTER 9

9.6.1 The capacitance tuning ratio of a varactor is obtained by calculating the maximum capacitance, at $\phi/2$, and the minimum value at -50 V. Taking the ratio of the two yields

$$\frac{C_{\text{max}}}{C_{\text{min}}} = \left(\frac{1 + 50/0.7}{1 - 1/2}\right)^{\gamma} = 144.85^{\gamma}.$$

For $\gamma = 0.5$, the ratio becomes 12, for $\gamma = 0.35$ it is reduced to 5.7, while with $\gamma = 0.25$ it drops to 3.47.

9.6.2   Considering Eq. 9.5, we must solve equation

$$\frac{(\alpha U_1)^4}{64} = 0.05 \frac{(\alpha U_1)^2}{4},$$

which provides the value of U_1:

$$U_1 = \frac{1}{\alpha} \sqrt{\frac{64}{4} 0.05} = \frac{0.894}{\alpha}.$$

References

Aanandan, C. K., and K. G. Nair (1986), "Compact Broadband Microstrip Antenna," *Electron. Lett.* **22**:1064–1065.

Abouzhara, M. D., and K. C. Gupta (1988), "Multiport Power Divider-Combiner Circuits Using Circular-Sector-Shaped Planar Components," *IEEE Trans. Microwave Theory Tech.* **32**(12):1747–1751.

Abrie, P. L. (1986), *The Design of Impedance-Matching Networks for Radio-Frequency and Microwave Amplifiers*. Norwood, MA: Artech House.

Agrawal, A. K., and G. F. Mikucki (1986), "A Printed Circuit Hybrid Ring Directional Coupler for Arbitrary Power Divisions," *IEEE Trans. Microwave Theory Tech.* **34**(12):1401–1407.

Akhtarzad, S., and P. B. Johns (1974), "Numerical Solution of Lossy Waveguides: T. L. M. Computer Program," *Electron. Lett.* **10**(10):309–311.

Akhtarzad, S., and P. B. Johns (1975a), "Three-Dimensional Transmission-Line Matrix Computer Analysis of Microstrip Resonators," *IEEE Trans. Microwave Theory Tech.* **23**(12):990–997.

Akhtarzad, S., and P. B. Johns (1975b), "Dispersion Characteristics of a Microstrip Line with a Step Discontinuity," *Electron. Lett.* **11**:310–311.

Akhtarzad, S., T. R. Rowbotham, and P. B. Johns (1975), "The Design of Coupled Microstrip Lines," *IEEE Trans. Microwave Theory Tech.* **23**(6):486–492.

Alanen, E., I. V. Lindell, and A. T. Hujanen (1986), "Exact Image Method for Field Calculation in Horizontally Layered Medium above a Conducting Ground Plane," *IEE Proc.–H* **133**(4):297–304.

Ali, F., and A. Gupta (1991), *HEMTs and HBTs: Devices, Fabrication and Circuits*, Norwood, MA: Artech House.

Anada, T., J.-P. Hsu, and T. Okoshi (1991), "New Synthesis Method for a Branch Line 3-dB Hybrid: A Hybrid Approach Comprising Planar and Transmission Line Circuit Concepts," *IEEE Trans. Microwave Theory Tech.* **39**(6):969–976.

Anders, P., and F. Arndt (1980), "Microstrip Discontinuity, Capacitances and Inductances for Double Steps, Mitered Bends with Arbitrary Angle, and Asymmetric Right-Angle Bends," *IEEE Trans. Microwave Theory Tech.* **28**:1213–1217.

Arabi, T. R., A. T. Murphy, T. K. Sarkar, R. F. Harrington, and A. R. Djordjevic (1991), "On the Modeling of Conductor and Substrate Losses in Multiconductor, Multidielectric Transmission Line Systems," *IEEE Trans. Microwave Theory Tech.* **39**:1090–1097.

Araki, K., and T. Itoh (1981), "Hankel Transform Domain Analysis of Open Circular Microstrip Radiating Structures," *IEEE Trans. Antennas Propag.* **25**:84–89.

Arndt, F. (1968), "High Pass Transmission Line Directional Coupler," *IEEE Trans. Microwave Theory Tech.* **16**:310.

Assadourian, F., and E. Rimai (1952), "Simplified Theory of Microstrip Transmission Systems," *Proc. IRE* **40**:1651–1657.

Ayasli, Y. (1989), "Field Effect Transistor Circulators," *IEEE Trans. Magnetics* **25**(5):3242–3247.

Bahl, I. J., and P. Bhartia (1980), *Microstrip Antennas*, Norwood, MA: Artech House.

Baiocchi, O. R., K.-S. Kong, and T. Itoh (1992), "Pulse Propagation in Superconducting Coplanar Striplines," *IEEE Trans. Microwave Theory Tech.* **40**:509–514.

Beggs, J. H., R. J. Luebbers, K. S. Yee, and K. S. Kunz (1992), "Finite-Difference Time-Domain Implementation of Surface Impedance Boundary Conditions," *IEEE Trans. Antennas Propag.* **40**:49–56.

Benedek, P., and P. Silvester (1972), "Equivalent Capacitance for Microstrip Gaps and Steps," *IEEE Trans. Microwave Theory Tech.* **20**(11):729–733.

Besser, L. (1984), "High-Frequency CAD Comes out of the Lab and onto the Shelf," *Microwaves RF* **23**(12):65–69.

Bhattacharyya, A. K., and L. Shafai (1986), "Surface Wave Coupling between Circular Patch Antennas," *Electron. Lett.* **22**(22):1198–1200.

Bögelsack, F., and I. Wolff (1987), "Application of a Projection Method to a Mode-Matching Solution for Microstrip Lines with Finite Metallization Thickness," *IEEE Trans. Microwave Theory Tech.* **35**(10):918–921.

Bokhari, S. A., J. R. Mosig, and F. Gardiol (1992), "Radiation Pattern Computation of Microstrip Antennas on Finite Size Ground Plane," *IEE Proc.-H* **139**(3):278–286.

Bosma, H. (1962), "On the Principle of Waveguide Circulation," *Proc. IEE-B* **109**(21).

Boukamp, J., and R. H. Jansen (1984), "The High-Frequency Behavior of Microstrip Open-Ends in Microwave Integrated Circuits Including Energy Leakage," *Proc. Fourteenth European Microwave Conf.*, pp. 142–147.

Bourreau, D., B. Della, E. Daniel, C. Person, and S. Toutain (1992), "High Performance Lange Coupler," *Electron. Lett.* **28**(21):1997–1998.

Bracewell, R. N. (1978), *The Fourier Transform and Its Applications*, New York: McGraw-Hill.

Brodie, I., and J. J. Muray (1992), *The Physics of Micro/Nano Fabrication*, New York: Plenum.

Bryant, T. J., and J. A. Weiss (1968), "Parameters of Microstrip Transmission Lines and of Coupled Pairs of Transmission Lines," *IEEE Trans. Microwave Theory Tech.* **16**(12):1021–1027.

Carroll, J. E. (1970), *Hot Electron Microwave Generators*, London: Edward Arnold.

Carver, K. R. (1979), *Proceedings of the Workshop on Printed Antenna Technology*, New Mexico State University, Las Cruces, NM.

Carver, K. R., and J. W. Mink (1981), "Microstrip Antenna Technology," *IEEE. Trans. Antennas Propag.* **29**:2–24.

Chadha, R. (1981), *Triangular Segments and Two-Dimensional Analysis for Microwave Integrated Circuits*, Ph.D. thesis, Indian Institute of Technology, Kanpur, India.

Chadha, R., and K. C. Gupta (1982), "Compensation of Discontinuities in Planar Transmission Lines," *IEEE Trans. Microwave Theory Tech.* **30**:2151–2156.

Chang, D. C. (1981), Special Issue of *IEEE AP-S Transactions on Antennas and Propagation* **29**(1).

Chang, K. (1990), *Handbook of Microwave and Optical Components*, vol. 2, New York: Wiley.

Chang, K., T. S. Martin, F. Wang, and L. C. Klein (1987), "On the Study of Microstrip Ring and Varactor Tuned Ring Circuits," *IEEE Trans. Microwave Theory Tech.* **35**(12):1288–1295.

Choi, D. H., and W. J. R. Hoefer (1986), "The Finite-Difference–Time-Domain Method and Its Application to Eigenvalue Problems," *IEEE Trans. Microwave Theory Tech.* **34**(12):1464–1470.

Chu, T. S., T. Itoh, and Y.-C. Shih (1985), "Comparative Study of Mode-Matching Formulations for Microstrip Discontinuity Problems," *IEEE Trans. Microwave Theory Tech.* **33**(10):1018–1023.

Chuang, S. L., L. Tsang, J. A. Kong, and W. C. Chew (1980), "The Equivalence of the Electric and Magnetic Surface Current Approaches in Microstrip Antenna Studies, *IEEE Trans. Antenna Propag.* **28**:569–571.

Cooper, W., G. B. Norris, and C. Barratt (1989), "High Yield, 0.4 W, 2–18 GHz GaAs Distributed Amplifiers," *Appl. Microwave* **1**:98–106.

Costache, G. I. (1987), "Finite Element Method Applied to Skin-Effect Problems in Strip Transmission Lines," *IEEE Trans. Microwave Theory Tech.* **35**(11):1009–1013.

Cristal, E. G., and S. Frankel (1972), "Hairpin-Line and Hybrid Hairpin-Line/Half Wave Parallel Coupled Filters," *IEEE Trans. Microwave Theory Tech.* **20**(11):719–728.

Csendes, Z. J. (1991), "Finite Element Modeling of Microwave and MM-wave Discontinuities," presented at the International Workshop of the German IEEE/AP Chapter, *CAD Oriented Numerical Techniques for the Analysis of Microwave and MM-Wave Transmission-Line Discontinuities and Junctions*, Stuttgart, Germany.

Csendes, Z. J., and J.-F. Lee (1988), "The Transfinite Element Method for Modeling MMIC Devices," *IEEE Trans. Microwave Theory Tech.* **36**(12):1639–1649.

Daly, P. (1971), "Hybrid-Mode Analysis of Microstrip by Finite-Element Methods," *IEEE Trans. Microwave Theory Tech.* **19**(1):19–25.

Das, N. K., and D. M. Pozar (1989), "Analysis and Design of Series-fed Arrays of Printed-Dipoles Proximity-Coupled to a Perpendicular Microstripline," *IEEE Trans. Antenna Propag.* **37**(4):435–444.

Davies, J. B. (1989), "The Finite Element Method," in *Numerical Techniques for Microwave and Millimeter-Wave Passive Structures*, T. Itoh, Ed., New York: Wiley, pp. 33–132.

Demeure, L. (1986), "New Low Cost and Low Loss Substrate: Application to Printed Antenna," in *Proceedings Journées Internationales de Nice sur les Antennes (JINA)*.

Denig, C. (1989), "Using Microwave CAD Programs to Analyze Microstrip Interdigital Filters," *Microwave J.* **32**(3):147–152.

Derneryd, A. G. (1978), "A Theoretical Investigation of the Rectangular Microstrip Antenna Element," *IEEE Trans. Antennas Propag.* **26**:532–535.

Derneryd, A. (1988), "New Technologies for Lightweight, Low Cost Antennas," in *Workshop on Lightweight Antennas, Eighteenth European Microwave Conference*, Stockholm, Sweden.

Deschamps, G. A. (1953), "Microstrip Microwave Antennas," in *Third USAF Symposium on Antennas*.

Dobrowolski, J. A. (1991), *Introduction to Computer Methods for Microwave Circuit Analysis and Design*, Norwood, MA: Artech House.

Douville, R. J. P., and D. S. James (1978), "Experimental Study of Symmetric Microstrip Bends and Their Compensation," *IEEE Trans. Microwave Theory Tech.* **26**:175–181.

Dubost, G. (1981), *Flat Radiating Dipoles and Their Application to Arrays*, New York: Wiley.

Dunleavy, L. P., and P. B. Katehi (1988), "A Generalized Method for Analyzing Shielded Thin Microstrip Discontinuities," *IEEE Trans. Microwave Theory Tech.* **36**(12):1758–1766.

Dydyk, M. (1977), "Master the T-Junction and Sharpen Your MIC Design," *MicroWaves* **5**:184–186.

Dydyk, M. (1986), "Planar Radial Resonator Oscillator," *Digest IEEE MTT-S Int. Microwave Symp.* pp. 167–168.

Easter, B. (1975), "The Equivalent Circuit of some Microstrip Discontinuities," *IEEE Trans. Microwave Theory Tech.* **23**:655–660.

El-Ghazaly, S. M., R. B. Hammond, and T. Itoh (1992), "Analysis of Superconducting Structures: Application to Microstrip Lines," *IEEE Trans. Microwave Theory Tech.* **40**:499–508.

Engen, G., and R. W. Beatty (1959), "Microwave Reflectometer Techniques," *IRE Trans. Microwave Theory Tech.* **7**:351–355.

Engen, G., and C. A. Hoer (1979), "Thru-Reflect-Line: An Improved Technique for Calibrating the Dual Six-Port Automatic Network Analyzer," *IEEE Trans. Microwave Theory Tech.* **27**(12):987–993.

Farrar, A., and A. T. Adams (1972), "Matrix Method for Microstrip Three-Dimensional Problems," *IEEE Trans. Microwave Theory Tech.* **20**:497–504.

Fay, C. E., and R. L. Comstock (1965), "Operation of the Ferrite Junction Circulator," *IEEE Trans. Microwave Theory Tech.* **13**(1):15–27.

Feix, N., M. Lalande, and B. Jecko (1992), "Harmonical Characterization of a Microstrip Bend via the Finite Difference Time Domain Method," *IEEE Trans. Microwave Theory Tech.* **40**(5):955–961.

Forsythe, G. F., and W. R. Wasow (1960), *Finite Difference Methods for Partial Differential Equations*, New York: Wiley.

Franzen, N. R., and R. A. Speciale (1975), "A New Procedure for System Calibration and Error Removal in Automated S-Parameter Measurements," in *Proceedings of the Fifth European Microwave Conference*, Hamburg, Germany.

Friis, H. T. (1967), "Analysis of Harmonic Generator Circuits for Step-Recovery Diodes," *Proc. IEEE* **57**:1192–1194.

Gardiol, F. (1984), *Introduction to Microwaves*, Norwood, MA: Artech House.

Gardiol, F. (1987), *Lossy Transmission Lines*, Norwood, MA: Artech House.

Gardiol, F., guest editor (1992), "Special Issue on Electromagnetic Simulation of Planar Microwave and Millimeter Wave Circuits," *Int. J. Microwave/Millimeter Wave Computer-Aided Eng.* **2**(4).

Garg, R., and I. J. Bahl (1978), "Microstrip Discontinuities," *Int. J. Electron.* **1**:81–87.

Getsinger, W. J. (1973), "Microstrip Dispersion Model," *IEEE Trans. Microwave Theory Tech.* **MTT-21**:34–39.

Gibbons, G. (1973), *Avalanche Diode Microwave Oscillators*, Oxford: Clarendon Press.

Golio, J. M. (1991), *Microwave MESFETs und HEMTs*, Norwood, MA: Artech House.

Gopinath, A., A. F. Thomson, and I. M. Stephenson (1976), "Equivalent Circuit Parameter of Microstrip Step Change in Width and Cross Sections," *IEEE Trans. Microwave Theory Tech.* **24**:142–144.

Grieg, D. D., and H. F. Engelmann (1952), "Microstrip—A New Transmission Technique for the Kilomegacycle Range," *Proc. IRE* **40**:1644–1650.

Grünberger, G. K., and H. H. Meinke (1971), "Experimenteller und theoretischer Nachweis der Längsfeldstärken in der Grundwelle der Mikrowellen-Streifenleitung," *Nachr. Z.* **24**:364–368.

Guglielmi, M., and J. M. Fernandez (1990), "Dual Mode Filters Based on Microstrip Resonators," *Microwave Eng. Europe* Sept./Oct.:41–49.

Gunn, J.-B. (1964), "Microwave Oscillations of Current in III-V Semiconductors," *IBM J. Res. Dev.* **8**:103–106.

Gunston, M. A. R. (1972), *Microwave Transmission Line Impedance Data*, New York: Van Nostrand Reinhold.

Gunton, D. J. (1978), "Design of Wideband Codirectional Couplers and Their Realization at Microwave Frequencies Using Coupled Comblines," *IEE Microwaves, Optics Acoustics* **2**:19–30.

Gupta, K. C. (1989), "Planar Circuit Analysis," in *Numerical Techniques for Microwave and Millimeter-Wave Passive Structures*, T. Itoh, Ed., New York: Wiley, pp. 214–333.

Gupta, K. C., and M. D. Abouzhara (1985), "Analysis and Design of Four-Port and Five-Port Microstrip Disc Circuits," *IEEE Trans. Microwave Theory Tech.* **33**(12):1422–1428.

Gupta, K. C., R. Garg, and I. J. Bahl (1979), *Microstrip Lines and Slotlines*, Norwood, MA: Artech House.

Gupta, K. C., R. Garg, and R. Chadha (1981), *CAD of Microwave Circuits*, Norwood, MA: Artech House.

Gwarek, W. K. (1988), "Analysis of Arbitrarily Shaped Two-dimensional Microwave Circuits by Finite-Difference Time-Domain," *IEEE Trans. Microwave Theory Tech.* **36**(4):738–744.

Hafner, C. (1990), *The Generalized Multipole Method for Computational Electromagnetics*, Norwood, MA: Artech House.

Hall, P. S., and C. M. Hall (1988), "Coplanar Corporate Feed Effects in Microstrip Patch Array Design," *IEE Proc.-H* **135**(3):180–186.

Hall, R. C., and J. R. Mosig (1989), "Vertical Monopoles Embedded in a Dielectric Substrate," *IEE Proc. (London)-H* **136**(6):462–468.

Hammerstad, E. O. (1975), "Equations for Microstrip Circuit Design," in *Proceedings of the European Microwave Conference*, Hamburg, Germany, pp. 268–272.

Hammerstad, E. O., and F. Bekkadal (1975), *Microstrip Handbook*, Trondheim: Norwegian Institute of Technology, report ELAB STF44 A74169.

Hammerstad, E. O., and O. Jensen (1980), "Accurate Models for Microstrip Computer-Aided Design," *IEEE MTT-S Digest Int. Microwave Symp. USA*, pp. 407–409.

Hansen, R., Ed. (1966), *Microwave Scanning Antennas*, New York: Academic Press.

Hara, S., T. Tokumitsu, and M. Aikawa (1990), "Novel Unilateral Circuits for MMIC Circulators," *IEEE Trans. Microwave Theory Tech.* **38**(10):1399–1406.

Harrington, R. F. (1968), *Field Computation by Moment Methods*, New York: Macmillan.

Haupt, G., and H. Delfs (1974), "High-Directivity Microstrip Directional Couplers," *Electron. Lett.* **10**:142–143.

Hearn, C. P., E. S. Bradshaw, and R. J. Trew (1990), "The Effect of Coupling Line Loss in Microstrip to Dielectric Resonator Coupling," *Microwave J.* **33**(11):169–172.

Hechtman, C., H. Zmuda, and D. Gabbay (1991), "A Frequency-Dependent Basis Function Applied to Microstrip," *IEEE Trans. Microwave Theory Tech.* **39**(5):893–896.

Herzog, H.-J. (1978), "Microstrip Couplers with Improved Directivity," *Electron. Lett.* **14**:50–51.

Hewlett-Packard (1992), "Network Analysis: Applying the HP 8510 TRL Calibration for Non-coaxial Measurements," Product note 8510-8A.

Hill, A. (1991), "Analysis of Multiple Coupled Microstrip Discontinuities for Microwave and Millimeter Integrated Circuits," *Digest of IEEE MTT-S*, Symposium pp. 1091–1094.

Hill, A., J. Burke, and K. Kottapalli (1992), "Three Dimensional Electromagnetic Analysis of Shielded Microstrip Circuits," *Int. J. Microwave/Millimeter Wave Computer-Aided Eng.* **2**(4):286–296.

Hill, A., and V. K. Tripathi (1991), "An Efficient Algorithm for the Three-Dimensional Analysis of Passive Microstrip Components and Discontinuities for Microwave and Millimeter-wave Integrated Circuits," *IEEE Trans. Microwave Theory Tech.* **39**(1):83–91.

Hoefer, W. J. R. (1977), "Equivalent Series Inductivity of a Narrow Transverse Slit in Microstrip," *IEEE Trans. Microwave Theory Tech.* **25**:822–824.

Hoefer, W. J. R. (1989), "The Transmission Line Matrix (TLM) Method," in *Numerical Techniques for Microwave and Millimeter Wave Passive Structures*, T. Itoh, Ed., New York: Wiley, pp. 496–591.

Hoefer, W. J. R., and A. Chattopadhyay (1975), "Evaluation of the Equivalent Circuit Parameters of Microstrip Discontinuities through Perturbation of a Resonant Ring," *IEEE Trans. Microwave Theory Tech.* **23**(12):1067–1071.

Hoefer, W. J. R., and Y.-C. Shih, (1980), "The Accuracy of TLM Analysis of Finned Rectangular Waveguides," *IEEE Trans. Microwave Theory Tech.* **28**(7):743–746.

Hoffmann, R. K. (1987), *Handbook of Microwave Integrated Circuits*, Norwood, MA: Artech House.

Hofmann, G. R. (1984), "Introduction to Computer Aided Design of Microwave Circuits," in *Proceedings of the 14th European Microwave Conference*, Liège, Belgium, pp. 731–737.

Howard, P. J. (1992), "New Microwave and mm-Wave Substrates Using Glass-Ceramic Technology," *Microwave J.* **35**(5):274–279.

Howes, M. J., and D. V. Morgan (1976), *Microwave Devices, Device Circuit Interactions*, London: Wiley.

Howes, M. J., and D. V. Morgan (1978), *Variable Impedance Devices*, New York: Wiley.

Hygate, G., and J. F. Nye (1990), "Measuring Microwave Fields Directly with an Optically Modulated Scatterer," *Meas. Sci. Technol.* **1**:703–709.

Hygate, G., and J. F. Nye (1991), "Measuring a Microwave Field Close to a Conductor," *Meas. Sci. Technol.* **2**:838–845.

IPC (1988), "Stripline Test for Permittivity and Loss Tangent at X-band," Institute for Interconnections and Packaging Electronics Circuits, IPC-TM-650, Method 2.5.5.5.

Ishimaru, A., R. G. Coe, G. E. Miller, and W. P. Geren (1985), "Finite Periodic Structure Approach to Large Scanning Array Problems," *IEEE Trans. Antennas Propag.* **33**:1213–1220.

Islam, S. (1988), "The Design of Microwave Forward Directional Couplers Using Microstrip Comblines," *Microwave J.* **31**(11):83–100.

Itoh, T. (1980), "Spectral Domain Immitance Approach for Dispersion Characteristics of Generalized Printed Transmission Lines," *IEEE Trans. Microwave Theory Tech.* **29**(7):496–499.

Itoh, T., Ed. (1989), *Numerical Techniques for Microwave and Millimeter Wave Passive Structures*, New York: Wiley.

Itoh, T., and R. Mittra (1973a), "A New Method for Calculating the Capacitance of a Circular Disk for Microwave Integrated Circuits," *IEEE Trans. Microwave Theory Tech.* **21**(6):431–432.

Itoh, T., and R. Mittra (1973b), "Spectral-domain Approach for Calculating the Dispersion Characteristics of Microstrip Lines," *IEEE Trans. Microwave Theory Tech.* **21**(7):733–736.

Jackson, C. M. (1989), "Introduction to Lange Coupler Design," *Microwave J.* **32**(10):145–149.

Jackson, D. R., and A. Oliner (1988), "A Leaky Wave Analysis of the High-Gain Printed Antenna Configuration," *IEEE Trans. Antennas Propag.* **36**(7):905–910.

Jackson, R. W. (1989), "Full-Wave, Finite Element Analysis of Irregular Microstrip Discontinuities," *IEEE Trans. Microwave Theory Tech.* **37**:81–89.

Jackson, R. W., and D. M. Pozar (1985), "Full-Wave Analysis of Microstrip Open-end and Gap Discontinuities," *IEEE Trans. Microwave Theory Tech.* **33**:1036–1042.

James, D. S., and H. S. Tse (1972), "Microstrip End Effects," *Electron. Lett.* **8**:46–47.

James, J. R., and A. Henderson (1979), "High-Frequency Behavior of Microstrip Open-Circuit Terminations," *IEE J. Microwaves, Optics Acoustics*, **3**(5):205–218.

James, J. R., P. S. Hall, and C. Wood (1981), *Microstrip Antenna Theory and Design*, London: IEE Press.

Jansen, R. H. (1974), "Shielded Rectangular Microstrip Disc Resonators," *Electron. Lett.* **10**(15):299–300.

Jansen, R. H. (1981), "Hybrid Mode Analysis of End Effects of Planar Microwave and Millimetre Wave Transmission Lines," *IEE Proc.-H* **128**(2):77–86.

Jansen, R. H. (1985), "The Spectral Domain Approach for Microwave Integrated Circuits," *IEEE Trans. Microwave Theory Tech.* **33**:1043–1056.

Jansen, R. H. (1986), "LINMIC, a CAD Package for the Layout-Oriented Design of Single and Multilayer MICs/MMICs up to mm Wave Frequencies," *Microwave J.* **29**(2):151–161.

Jansen, R. H. (1988), "A Comprehensive CAD Approach to the Design of Monolithic Microwave Integrated Circuits," *IEEE Trans. Microwave Theory Tech.* **36**:208–219.

Janssen, W. (1977), *Hohlleiter und Streifenleiter*, Heidelberg: Hüthig.

Johns, P. B., and S. Akhtarzad (1975), "Dispersion Characteristic of a Microstrip Line with a Step Discontinuity," *Electron. Lett.* **11**(14):310–311.

Johns, P. B., and R. L. Beurle (1971), "Numerical Solution of 2-Dimensional Scattering Problems Using a Transmission-Line Matrix," *IEE Proc.* **118**(9):1203–1208.

Johnson, R. A. (1975), "Understanding Microwave Power Splitters," *Microwave J.* **18**(12):49–56.

Julien, N., and J. C. Mollier (1992), "Direct Synthesis of Distributed Lossless Networks with Application to Microwave Amplifier Design," *Int. J. Microwave Millimeter Wave Computer Aided Eng.* **2**(2):70–81.

Jung, W.-L., and J. Wu (1990), "Stable Broadband Microwave Amplifier Design," *IEEE Trans. Microwave Theory Tech.* **38**(8):1079–1085.

Kajfez, D., and M. D. Tew (1980), "Pocket Calculator Program for Analysis of Lossy Microstrip," *Microwave J.* **23**(12):39–48.

Katehi, P. B., and N. G. Alexopoulos (1984), "On the Modelling of Electromagnetically Coupling Microstrip Antennas," *IEEE Trans. Antennas Propag.* **32**:1179–1186.

Katehi, P. B., and N. G. Alexopoulos (1985), "Frequency-dependent Characteristics of Microstrip Discontinuities in Millimeter-Wave Integrated Circuits," *IEEE Trans. Microwave Theory Tech.* **33**(10):1029–1035.

Katzin, P., Y. Ayasli, L. D. Reynolds, and B. E. Bedard (1992), "6–18 GHz MMIC Circulators," *Microwave J.* **35**(5):248–256.

Kirschning, M., and R. H. Jansen (1982), "Accurate Model for Effective Dielectric Constant of Microstrip with Validity up to Millimeter Wave Frequencies," *Electron. Lett.* **18**:272–273.

Kirschning, M., and R. H. Jansen (1984), "Accurate Wide-Range Design Equations for the Frequency-Dependent Characteristic of Parallel-Coupled Microstrip Lines," *IEEE Trans. Microwave Theory Tech.* **32**:83–90.

Kirschning, M., R. H. Jansen, and N. L. H. Koster (1981), "Accurate Model for Open-Ended Effect of Microstrip Lines," *Electron. Lett.* **17**(3):123–125.

Kirschning, M., R. Jansen, and N. H. L. Koster (1983), "Measurement and Computer-Aided Modelling of Microstrip Discontinuities by an Improved Resonator Method," in *Digest IEEE S-MTT International Microwave Symposium*.

Kishk, A. A., and L. Shafai (1986a), "Different Formulations for Numerical Solution of Single or Multibodies of Revolution with Mixed Boundary Conditions," *IEEE Trans. Antennas Propag.* **34**(5):666–673.

Kishk, A. A., and L. Shafai (1986b), "The Effect of Various Parameters of Circular Microstrip Antennas on their Radiation Efficiency and the Mode Excitation," *IEEE Trans. Antennas Propag.* **34**(8):969–976.

Klein, J. L., and K. Chang (1990), "Optimum Dielectric Overlay Thickness for Equal Even- and Odd-Mode Phase Velocities in Coupled Microstrip Circuits. *Electron. Lett.*, **26**(5):274–276.

Koike, S., N. Yoshida, and I. Fukai (1985), "Transient Analysis of Microstrip Gap in Three-Dimensional Space," *IEEE Trans. Microwave Theory Tech.* **33**(8):726–730.

Kollberg, E., Ed. (1984), *Microwave and Millimeter Wave Mixers*, New York: IEEE Press.

Kollberg, E. (1990), "Mixers and Detectors," in *Handbook of Microwave and Optical Components*, K. Chang, Ed., vol. 2, New York: Wiley.

Kompa, G. (1976a), "S-Matrix Computations of Microstrip Discontinuities with a Planar Waveguide Model," *Arch. Elektr. Übertragungstech.* **30**:58–64.

Kompa, G. (1976b), "Reduced Coupling Aperture of Microstrip Stubs Provides New Aspects in Stub Filter Design," in *Proceedings of the Sixth European Microwave Conference*, Rome, pp. 39–43.

Konrad, A. (1977), "Higher-Order Triangular Finite Elements for Electromagnetic Waves in Anisotropic Media," *IEEE Trans. Microwave Theory Tech.* **25**(5):353–360.

Koster, H. L., and R. H. Jansen (1986), "The Microstrip Step Discontinuity: A Revisited Description," *IEEE Trans. Microwave Theory Tech.* **34**(2):213–223.

Ladbrooke, P. H., M. Potok, and E. H. England (1973), "Coupling Errors in Cavity Resonance Measurements on MIC Dielectrics," *IEEE Trans. Microwave Theory Tech.* **21**:560–562.

Lange, J. (1979), "Interdigited Stripline Quadrature Hybrid," *IEEE Trans. Microwave Theory Tech.* **17**:1150–1151.

Laverghetta, T. S. (1991), *Microwave Materials and Fabrication Techniques*, Norwood, MA: Artech House.

Lee, D. L. (1986), *Electromagnetic Principles of Integrated Optics*, New York: Wiley.

Leighton, W., and A. G. Milnes (1971), "Junction Reactance and Dimensional Tolerance Effects on X-Band 3-dB Directional Couplers," *IEEE Trans. Microwave Theory Tech.* **19**:814–824.

Lepeltier, P., J. M. Floch, and J. Citerne (1985), "Electromagnetically Coupled Microstrip Dipole, Analysis by Means of Integral Equations," in *IEEE AP-S International Symposium Digest*, Vancouver, pp. 777–780.

Levine, E. (1989), "Special Measurement Techniques for Printed Antennas," in *Handbook of Microstrip Antennas*, J. R. James and P. S. Hall, Eds., vol. 2, chap. 16, London: Peter Peregrinus.

Levy, R. (1963), "General Synthesis of Asymmetric Multi-element Directional Couplers," *IEEE Trans. Microwave Theory Tech.* **11**:226–237.

Levy, R. (1964), "Tables of Asymmetric Multi-Element Coupled Transmission Line Directional Coupler," *IEEE Trans. Microwave Theory Tech.* **12**:275–279.

Li, C. Q., S. H. Li, and R. G. Bosisio (1984), "CAE-CAD Design of an Improved Wideband Wilkinson Power Divider," *Microwave J.*, **27**:125–135.

Liechti, C. (1989), "High-Speed Transistors: Directions for the 1990s," *Microwave J.* Special Issue: 1989 State of the Art Reference. **32**, pp. 165–177.

Lier, L. (1982), "Improved Formulas for Input Impedance of Coax-Fed Microstrip Patch Antennas," *Proc. IEE-H (MOA)* **129**:161–164.

Lo, Y. T., D. Solomon, and W. F. Richards (1979), "Theory and Experiment on Microstrip Antennas," *IEEE Trans. Antennas Propag.* **27**:137–145.

Maas, S. A. (1986), *Microwave Mixers*, Norwood, MA: Artech House.

Maeda, M. (1972), "An Analysis of Gap in Microstrip Transmission Line," *IEEE Trans. Microwave Theory Tech.* **20**:390–396.

Malherbe, J. A. G. (1979), *Microwave Transmission Line Filters*, Norwood, MA: Artech House.

Maloney, J. G., and G. S. Smith (1992), "The Use of Surface Impedance Concepts in the Finite-difference Time-Domain Method," *IEEE Trans. Antennas Propag.* **40**:38–48.

Manley, J. M., and H. E. Rowe (1956), "Some General Properties of Nonlinear Elements," *Proc. IRE* **44**(6):904–913.

March, S. L. (1984), "Microwave Circuit Layout: A Dynamic Plot Emerges," *Microwaves RF* **23**(12):59–161.

Marcuvitz, N. (1951), *Waveguide Handbook*, New York: McGraw-Hill.

Marsden, J. E. (1973), *Basic Complex Analysis*, San Francisco: W. H. Freeman.

Matick, R. E. (1969), *Transmission Lines for Digital and Communications Networks*, New York: McGraw-Hill.

Matthaei, G. L., L. Young, and E. M. T. Jones (1964), *Microwave Filters, Impedance-Matching Networks and Coupling Structures*, chap. 13, New York: McGraw-Hill.

Mehran, R. (1975), "The Frequency-dependent Scattering Matrix of Microstrip Right-Angle Bends, T-junctions and Crossings," *Arch. Elektr. Ubertragungstech.* **29**:454–460.

Mehran, R. (1976), "Frequency Dependent Equivalent Circuits for Microstrip Right-Angle Bends, T-junctions and Crossings," *Arch. Elektr. Ubertragungstech.* **30**:80–82.

Menzel, W. (1976), "Frequency-Dependent Transmission Properties of Truncated Microstrip Right-Angle Bends," *Electron. Lett.* **12**:641.

Menzel, W. (1978), "Frequency-Dependent Transmission Properties of Microstrip Y-junctions and 120° Bends," *IEE Microwaves, Optics and Acoustics (MOA)* **2**:55–59.

Menzel, W., and I. Wolff (1977), "A Method for Calculating the Frequency Dependent Properties of Microstrip Discontinuities," *IEEE Trans. Microwave Theory Tech.* **25**:107–112.

Mikucki, G. F., and A. K. Agrawal (1989), "A Broad-band Printed Circuit Hybrid Ring Power Divider," *IEEE Trans. Microwave Theory Tech.* **37**(1):112–117.

Mongia, R. K. (1990), "Resonant Frequency of Cylindrical Dielectric Resonator Placed in an MIC Environment," *IEEE Trans. Microwave Theory Tech.* **38**(6):802–804.

Montgomery, C. G., R. H. Dicke, and E. M. Purcell (1948), *Principles of Microwave Circuits*, New York: McGraw-Hill.

Moore, J., and H. Ling (1990), "Characterization of a 90° Microstrip Bend with Arbitrary Miter via the Time-Domain Finite Difference Method," *IEEE Trans. Microwave Theory Tech.* **38**(4):405–410.

Morse, P. M., and H. Feshbach (1953), *Methods of Theoretical Physics*, New York: McGraw-Hill.

Mosig, J. R. (1988), "Arbitrarily Shaped Microstrip Structures and Their Analysis with a Mixed Potential Integral Equation," *IEEE Trans. Microwave Theory Tech.* **36**:314–323.

Mosig, J. R., and F. Gardiol (1982), "A Dynamical Radiation Model for Microstrip Structures," in *Advances in Electronics and Electron Physics*, P. Hawkes, Ed., vol. 59, New York: Academic Press.

Mosig, J. R., and T. K. Sarkar (1986), "Comparison of Quasi-static and Exact Electromagnetic Fields from a Horizontal Electric Dipole above a Lossy Dielectric Backed by an Imperfect Ground Plane," *IEEE Trans. Microwave Theory Tech.* **34**:379–387.

Mosig, J. R., R. L. Hall, and F. Gardiol (1989), "Numerical Analysis of Microstrip Patch Antennas," in *Handbook of Microstrip Antennas*, J. R. James and P. S. Hall, Eds., vol. 1, chap. 8, London: Peter Peregrinus.

Mosig, J. R., and H. K. Smith (1992), "Printed Radiating Structures and Transitions in Multilayered Substrates," *Int. J. Microwave/Millimeter Wave Computer-Aided Eng.* **2**(4):273–285.

Munson, R. E. (1974), "Conformed Microstrip Antennas and Microstrip Phased Arrays," *IEEE Trans. Antennas Propag.* **22**:74–78.

Mur, G. (1981), "Absorbing Boundary Conditions for the Finite-Difference Approximation of the Time-Domain Electromagnetic-Field Equations," *IEEE Trans. Electromag. Compat.* **EMC-23**(4):377–382.

Murray, D. D. (1989), "Polynomial Expressions Lead to Accurate Resonant Frequency Computations for DROs," *Microwave J.* **32**(9):241–242.

Newman, E. H., and P. Tulyathan (1981), "Analysis of Microstrip Antennas Using the Method of Moments," *IEEE Trans. Antennas Propag.* **29**:47–53.

Ohkawa, S., K. Suyama, and H. Ishikawa (1975), "Low Noise Field Effect Transistors," *Fujitsu Sci. Tech. J.* **11**:151–173.

Okoshi, T., and T. Miyoshi (1972), "The Planar Circuit—An Approach to Microwave Integrated Circuitry," *IEEE Trans. Microwave Theory Tech.* **20**(4):245–252.

Owens, R. P. (1976), "Curvature Effects in Microstrip Ring Resonators," *Electron. Lett.* **12**:356–357.

Page, C. H. (1956), "Frequency Conversion with Positive Nonlinear Resistors," *J. Res. Nat. Bur. Standards* **56**, pp. 179–182.

Pantic, Z., and R. Mittra (1986), "Quasi-TEM Analysis of Microwave Transmission Lines by the Finite-Element Method," *IEEE Trans. Microwave Theory Tech.* **34**(11):1096–1103.

Parisot, M., and R. Soares (1988), "GaAs MESFETs: S-Parameter Measurements and Their Use in Circuit Design," in *GaAs Mesfet Circuit Design*, R. Soares, Ed., Norwood, MA: Artech House.

Paul, D. K., P. Gardner, and B.Y. Prasetyo (1991), "Broadband Branchline Coupler for S-band," *Electron. Lett.* **27**(15):1318–1319.

Pengelly, R. S. (1982), *Microwave Field Effect Transistors: Theory, Design and Applications*, New York: Wiley.

Pitzalis, O. (1989), "Microwave to mm-Wave CAE: Concept to Production," *Microwave J.* Special Issue: *1989* State of the Art Reference **32**:15–47.

Plonsey, R., and R. E. Collin (1961), *Principles and Applications of Electromagnetic Fields*, New York: McGraw-Hill.

Pozar, D. M. (1985a), "General Relations for a Phased Array of Printed Antennas Derived from Infinite Current Sheets," *IEEE Trans. Antennas Propag.* **33**:498–504.

Pozar, D. M. (1985b), "Microstrip Antenna Aperture-Coupled to a Microstrip-Line," *Electron. Lett.* **21**:49–50.

Pozar, D. M. (1989), "Analysis and Design Considerations for Printed Phased-Array Antennas," in *Handbook of Microstrip Antennas*, J. R. James and P. S. Hall, Eds., vol. 1, chap. 12, London: Peter Peregrinus.

Pozar, D. M., and D. H. Schaubert (1984), "Scan Blindness in Infinite Phased Arrays of Printed Dipoles," *IEEE Trans. Antennas Propag.* **32**(6):602–610.

Pucel, R. (1986), "MMICs, Modelling, and CAD—Where Do We Go from Here?," in *Proceedings of the 16th European Microwave Conference*, Dublin, Ireland, pp. 61–70.

Pucel, R., D. Massé, and C. P. Hartwig (1968), "Losses in Microstrip," *IEEE Trans. Microwave Theory Tech.* **16**:342–350.

Quirarte, J. L. R. (1991), "A Method for Designing Microwave Broadband Amplifiers by Using Chebyshev Filter Theory to Design the Matching Networks," *Microwave Opt. Tech. Lett.* **4**(3):117–123.

Railton, C. J., and J. P. McGeehan (1989), "Analysis of Microstrip Discontinuities Using the Finite Difference Time Domain Technique," *Digest IEEE MTT-S Symposium* 1009–1012.

Railton, C. J., and J. P. McGeehan (1990), "An Analysis of Microstrip with Rectangular and Trapezoidal Conductor Cross Sections," *IEEE Trans. Microwave Theory Tech.* **MTT-38**:1017–1022.

Railton, C. J., and S. A. Meade (1992), "Fast Rigorous Analysis of Shielded Planar Filters," *IEEE Trans. Microwave Theory Tech.* **40**(5):978–985.

Ramo, S., J. R. Whinnery, and T. Van Duzer (1965), *Fields and Waves in Communications Electronics*, New York: Wiley.

Rautio, J. C., and R. F. Harrington (1987), "An Electromagnetic Time-Harmonic Analysis of Shielded Microstrip Circuits," *IEEE Trans. Microwave Theory Tech.* **35**(8):726–730.

Reineix, A. K., and B. Jecko (1989), "Analysis of Microstrip Patch Antennas Using Finite Difference Time Domain Method," *IEEE Trans. Antennas Propag.* **37**(11):1361–1369.

Richmond, J. H. (1955), "A Modulated Scattering Technique for Measurement of Field Distributions," *IRE Trans. Microwave Theory Tech.* **3**:13–15.

Roduit, J.-C., A. Skrivervik, and J. F. Zürcher (1991), "Microstrip Compact Bandpass Filters," *Microwave Opt. Tech. Lett.*, **4**(10):384–387.

Rosloniec, S. (1990), *Algorithms for Computer-Aided Design of Linear Microwave Circuits*, Norwood, MA: Artech House.

Rubin, D. (1990), "De-embedding mm-Wave MICs with TRL," *Microwave J.* **33**(6):141–150.

Sagawa, M., K. Takabashi, and M. Makimoto (1989), "Miniaturized Hairpin Resonator Filters and Their Application to Receiver Front-End MIC's," *IEEE Trans. Microwave Theory Tech.* **37**(12):1991–1997.

Salerno, M., and R. Sorrentino (1986), "Planim: A New Concept in the Design of MIC Filters," *Electron. Lett.* **22**:1054–1056.

Salmer, G., R. Fauquembergue, M. Lefebvre, and A. Cappy (1988), "Modeling of Submicrometer Gate GaAs Field Effect Transistors," *Ann. Télécomm.* **43**(7–8):405–414.

Sanford, J., J. F. Zürcher, and S. Robert (1991), "Shaped Beam Patch Arrays for Mobile Communication Base Stations," *Microwave Engineering Europe* June/July:31–33.

Sarkar, T. K., and E. Arvas (1990), "An Integral Equation Approach to the Analysis of Finite Microstrip Antennas," *IEEE Trans. Antennas Propag.* **38**:305–312.

Schafer, M., U. Bochtler, R. Bitzer, and F. Landstorfer (1989), "Radiation Losses in Planar Filter Structures," *Microwave J.* **32**(10):139–143.

Schaller, G. (1977), "Optimization of Microstrip Directional Couplers with Lumped Capacitors," *Arch. Elektr. Ubertragungstechnik* **31**(7/8):301–307.

Schelkunoff, S. A. (1943), *Electromagnetic Waves*, New York: Van Nostrand Reinhold.

Schneider, M. V. (1969), "Microstrip Lines for Microwave Integrated Circuits," *Bell System Tech. J.* **48**:1421–1444.

Shafai, L., and A. A. Kishk (1989), "Analysis of Circular Microstrip Antennas," in *Handbook of Microstrip Antennas*, J. R. James and P. S. Hall, Eds., vol. 1, chap. 2, London: Peter Peregrinus.

Sharma, P. C., and K. C. Gupta (1981), "Desegmentation Method for Analysis of Two-dimensional Microwave Circuits," *IEEE Trans. Microwave Theory Tech.* **29**(10):1094–1098.

Sheen, D. M., S. M. Ali, M. D. Abouzahra, J. A. Kong (1990), "Application of the Three-dimensional Finite-Difference Time-Domain Method to the Analysis of Planar Microstrip Circuits," *IEEE Trans. Microwave Theory Tech.* **38**(7):849–857.

Shibata, T., T. Hayashi, and T. Kimura (1989), "Analysis of Microstrip Circuits Using Three-dimensional Full-wave Electromagnetic Field Analysis in the Time Domain," *IEEE Trans. Microwave Theory Tech.* **36**(6):1064–1070.

Shih, Y. C., and H. J. Kuno (1989), "Solid-State Sources from 1 to 100 GHz," *Microwave J.* Special Issue: 1989 State of the Art Reference **32**:145–161.

Silvester, P., and P. Benedek (1973), "Microstrip Discontinuity Capacitances for Right-Angle Bends, T-junctions and Crossings," *IEEE Trans. Microwave Theory Tech.* **21**:341–346.

Skrivervik, A. K., and J. R. Mosig (1990), "Impedance Matrix of Multiport Microstrip Discontinuities Including Radiation Effects," *Arch. Elektr. Ubertragungstechnik* **44**(6):453–461.

Skrivervik, A. K., and J. R. Mosig (1992), "Finite Phased Array of Microstrip Patch Antennas: The Infinite Array Approach," *IEEE Trans. Antennas Propag.* **40**:579–582.

Smith, Ph. (1969), *Electronic Applications of the Smith Chart in Waveguide, Circuit and Component Analysis*, New York: McGraw-Hill.

Smith, P. M., and A. W. Swanson (1989), "HEMTs—Low Noise and Power Transistors for 1 to 100 GHz," *Appl. Microwave* **1**:63–72.

Soares, R., Ed. (1988), *GaAs Mesfet Circuit Design*, Norwood, MA: Artech House.

Sommerfeld, A. (1909), "The Propagation of Waves in Wireless Telegraphy" (in German), *Ann. Phys.* (4)**28**:665.

Sorrentino, R. (1985), "Planar Circuits, Waveguide Models, and Segmentation Method," *IEEE Trans. Microwave Theory Tech.* **33**(10):1057–1066.

Sorrentino, R. (1989), *Numerical Methods for Passive Microwave and Millimeter Wave Structures*, New York: IEEE Press.

Spruth, W. G. (1989), *The Design of a Microprocessor*, Berlin: Springer-Verlag.

Sreenivas, A. I., and R. Stockton (1990), "Semiconductor Control Devices: Phase Shifters and Switches," in *Handbook of Microwave and Optical Components*, K. Chang, Ed., vol. 2, New York: Wiley.

Strang, G., and G. J. Fix (1973), *An Analysis of the Finite Element Method*, Englewood Cliffs, NJ Prentice-Hall.

Sucher, M., and J. Fox (1963), *Handbook of Microwave Measurements*, vol. 3, Brooklyn, NY: Polytechnic Press.

Sze, S. M. (1981), *Physics of Semiconductor Devices*, New York: Wiley.

Taflove, A., and M. E. Brodwin (1975a), "Numerical Solution of Steady-State Electromagnetic Scattering Problems Using the Time-Dependent Maxwell's Equations," *IEEE Trans. Microwave Theory Tech.* **23**(8):623–630.

Taflove, A., and M. E. Brodwin (1975b), "Computation of the Electromagnetic Fields and Induced Temperatures within a Model of the Microwave-Irradiated Human Eye," *IEEE Trans. Microwave Theory Tech.* **23**(11):888–896.

Taflove, A., and K. R. Umashankar (1989), "Review of FD-TD Numerical Modeling of Electromagnetic Wave Scattering and Radar Cross Section," *Proc. IEEE* **77**(5):682–699.

Tamura, H., T. Nishikawa, K. Wakino, and T. Sudo (1988), "Metalized MIC Substrates Using High K Dielectric Resonator Materials," *Microwave J.* **31**(10):117–126.

Thomson, A. F., and A. Gopinath (1975), "Calculation of Microstrip Discontinuity Inductances," *IEEE Trans. Microwave Theory Tech.* **20**(8):648–655.

Traut, G. R. (1989a), "Advances in Substrate Technology," in *Handbook of Microstrip Antennas*, J. R. James and P. S. Hall, Eds., vol. 2, chap. 15, London: Peter Peregrinus.

Traut, G. R. (1989b), "Thermoset Microwave Material Enhances Microwave Hybrids," *Appl. Microwave.* Spring Issue, **1**:80–86.

Twisleton, J. R. G. (1991), "The ABCD, Z and Y Matrix Description of Microwave Transistors," *Microwave J.* **34**(9):141–156.

Uenohara, M., and J. W. Gewartowski (1969), "Varactor Applications," in *Microwave Semiconductors and Their Circuit Applications*, H. A. Watson, Ed., pp. 228–258, New York: McGraw-Hill.

Uwano, T., R. Sorrentino, and T. Itoh (1987), "Characterization of Strip Line Crossings by Transverse Resonance Analysis," *IEEE Trans. Microwave Theory Tech.* **35**(12):1369–1376.

Uysal, S., and J. Watkins (1991), "Novel Microstrip Multifunction Directional Couplers and Filters for Microwave and Millimeter Wave Applications," *IEEE Trans. Microwave Theory Tech.* **39**:977–985.

Uzunoglu, N. K., C. N. Capsalis, and C. P. Chronopoulos (1988), "Frequency-Dependent Analysis of a Shielded Microstrip Step Discontinuity Using an Efficient Mode-Matching Technique," *IEEE Trans. Microwave Theory Tech.* **36**(6):976–984.

Van Bladel, J. (1975), "On the Resonances of a Dielectric Resonator of Very High Permittivity," *IEEE Trans. Microwave Theory Tech.* **23**:199–208.

Van Vleck, J. H. (1951), "The Relation between Absorption and Dispersion," in *Propagation of Short Radio Waves*, D. E. Kerr, Ed., New York: McGraw-Hill.

Vendelin, G. D. (1970), "Limitations on Stripline Q," *Microwave J.* **13**(5):63–69.

Wang, J., M. Spenuk, R. Fralich, and J. Litva (1991), "A Simple Approach to a Four-Way Hybrid Ring Power Divider Design," *Microwave and Opt. Tech. Lett.* **4**:61–64.

Watkins, J. (1969), "Circular Resonant Structures in Microstrip," *Electron. Lett.* **5**:524–525.

Watson, H. A., Ed. (1969), *Microwave Semiconductors and Their Circuit Applications*, pp. 228–258, New York: McGraw-Hill.

Waugh, R. W. (1989), "Sensitivity Analysis of the Lange Coupler," *Microwave J.* **32**(11):121–129.

Wheeler, H. (1965), "Transmission Line Properties of Parallel Strips Separated by a Dielectric Sheet," *IEEE Trans. Microwave Theory Tech.* **13**:172–185.

White, J. (1990), "Semiconductor Control Devices: PIN Diodes," in *Handbook of Microwave and Optical Components*, K. Chang, Ed., vol. 2, New York: Wiley.

Wilhelm, W. (1986), "Propagation Delays of Inter-connect Lines in Large Scale Integrated Circuits," *Siemens Forschung und Entwicklung Berichte* **15**(2):60–63.

Williams, D. A. (1991), "Millimeter-Wave Components and Subsystems Built Using Microstrip Technology," *IEEE Trans. Microwave Theory Tech.* **39**(5):768–774.

Wolff, I. (1989), "The Waveguide Model for the Analysis of Microstrip Discontinuities," in *Numerical Techniques for Microwave and Millimeter-Wave Passive Structures*, T. Itoh, Ed., chap. 7, New York: Wiley, pp. 447–495.

Wolff, I., G. Kompa, and R. Mehran (1972), "Calculation Method for Microstrip Discontinuities and T-junctions," *Electron. Lett.* **8**:177–179.

Wolff, I., and N. Knoppik (1971), "Microstrip Ring Resonators and Dispersion Measurement on Microstrip Lines," *Electron. Lett.* **7**:779–781.

Worm, S. B., and R. Pregla (1984), "Hybrid Mode Analysis of Arbitrarily Shaped Planar Microwave Structures by the Method of Lines," *IEEE Trans. Microwave Theory Tech.* **32**(2):191–196.

Wright, A. S., and S. K. Judah (1987), "Very Broadband Flat Coupling Hybrid Ring," *Electron. Lett.* **23**(1):47–49.

Wroblewski, R. (1988), "Synthesis and Design of Small-Signal and Low Noise GaAs MESFET Amplifiers," in *GaAs Mesfet Circuit Design*, R. Soares, Ed., Norwood, MA: Artech House.

Wu, D. I., D. C. Chang, and B. L. Brim (1991), "Accurate Numerical Modeling of Microstrip Junctions and Discontinuities," *Int. J. Microwave Millimeter-Wave Computer-Aided Eng.* **1**(1):48–58.

Wu, S.-C., H.-Y. Yang, N. Alexopoulos, and I. Wolff (1990), "A Rigorous Dispersive Characterization of Microstrip Cross and T-junctions," *IEEE Trans. Microwave Theory Tech.* **38**(12):1837–1990.

Wu, Y. S., and F. Rosenbaum (1973), "Mode Chart for Microstrip Ring Resonators," *IEEE Trans. Microwave Theory Tech.* **21**(12):487–489.

Xu, Y., and R. G. Bosisio (1992), "Calculation of Characteristic Impedances of Curved Microstrip Lines for MMIC and MHMIC," *Electron. Lett.* **28**(8):775–776.

Yamashita, E., Ed. (1990), *Analytical Methods for Electromagnetic Wave Problems*, Norwood, MA: Artech House.

Yang, H.-Y., J. A. Castaneda, and N. G. Alexopoulos (1992), "Surface Wave Modes of Printed Circuits on Ferrite Substrates," *IEEE Trans. Microwave Theory Tech.* **40**(4):613–621.

Yang, J. J., Y. L. Chow, G. E. Howard, and D. G. Fang (1992), "Complex Images of an Electric Dipole in Homogeneous and Layered Dielectrics between Two Ground Planes," *IEEE Trans. Microwave Theory Tech.* **40**(3):595–600.

Yee, H.-Y. (1985), "Transverse Modal Analysis of Printed Circuit Transmission Lines," *IEEE Trans. Microwave Theory Tech.* **33**(9):808–816.

Yee, H.-Y., and K. Wu (1986), "Printed Circuit Transmission-Line Characteristic Impedance by Transverse Modal Analysis," *IEEE Trans. Microwave Theory Tech.* **34**(11):1157–1163.

Yee, K. S. (1966), "Numerical Solution of Initial Boundary Value Problems Involving Maxwell's Equations in Isotropic Media," *IEEE Trans. Antennas Propag.* **14**(3):302–307.

Yeo, S. P. (1992), "Analysis of Symmetrical Six-Port Junction When Configured as a Six-Port Reflectometer," *IEEE Trans. Instrum. Meas.* **41**(2):193–197.

Yeo, S. P., T. S. Yeo, and K. M. Ng (1989), "First Order Eigenmode Analysis of Symmetrical Five-Port Microstrip Ring Coupler," *Microwave and Optical Tech. Lett.* **2**:91–94.

Yngvesson, S. (1991), *Microwave Semiconductor Devices*, Boston: Kluwer.

Zhang, X., J. Fang, K. K. Mei, and Y. Liu (1988), "Calculations of the Dispersive Characteristics of Microstrips by the Time-domain Finite Difference Method," *IEEE Trans. Microwave Theory Tech.* **36**(2):263–267.

Zhang, X., and K. K. Mei (1988), "Time-Domain Finite Difference Approach to the Calculation of the Frequency-Dependent Characteristics of Microstrip Discontinuities," *IEEE Trans. Microwave Theory Tech.* **36**(12):1775–1787.

Zürcher, B., J.-F. Zürcher, and F. Gardiol (1989), "Broadband Microstrip Radiators, the SSFIP Concept," *Electromagnetics* **9**:385–393.

Zürcher, J.-F. (1985), "MICROS 3—A CAD/CAM Program for Fast Realization of Microstrip Masks," in *Proceedings of the MTT-S International Microwave Symposium*, Saint Louis, pp. 481–484.

Zürcher, J.-F. (1988), "The SSFIP, a Global Concept for High Performance Broadband Planar Antennas," *Electron. Lett.* **24**(23):1433–1435.

Zürcher, J.-F. (1992), "Simple and Powerful Near Field Measurement Method Applied to Planar Structures," *Microwave Engineering Europe* **5**:43–51.

Zürcher, J. F., L. Barlatey, and F. Gardiol (1986), "Computer-Aided Method to Measure the Permittivity of Microstrip Substrates," in *Proceedings of MIOP Symposium*, Wiesbaden, Germany.

Zürcher, J.-F., and F. Gardiol (1989), "Computer-Aided Design of Microstrip and Triplate Circuits," in *Handbook of Microstrip Antennas*, J. R. James and P. S. Hall, Eds., vol. 2, chap. 17, London: Peter Peregrinus.

Zwamborn, A. P. M., and P. M. Van Den Berg (1991), "A Weak Form of the Conjugate Gradient Method for Plate Problems," *IEEE Trans. Antennas Propag.* **39**:224–229.

Index

Heterick Memorial Library
Ohio Northern University

DUE	RETURNED	DUE	RETURNED
1.		13.	
2.		14.	
3.		15.	
4.		16.	
5.		17.	
6.		18.	
7.		19.	
8.		20.	
9.		21.	
10.		22.	
11.		23.	
12.		24.	